The Open University

GW00598209

Mathematical Logic **Unit 8**

Gödel's Incompleteness Theorems

Prepared by the Course Team

This system proves its own consistency

$\forall x \, \forall y \, (x \cdot y) = (y \cdot x)$

$1+1=2$

$\mathbb{R}$ is not countable

This computation doesn't halt

The M381 Mathematical Logic Course Team

The Mathematical Logic half of the course was produced by the following team:

Roberta Cheriyan	*Course Manager*
Derek Goldrei	*Course Team Chair* and *Academic Editor*
Jeremy Gray	*History Consultant*
Mary Jones	*Critical Reader*
Roger Lowry	*Publishing Editor*
Alan Pears	*Author*
Alan Slomson	*Author*
Frances Williams	*Critical Reader*

with valuable assistance from:

The Maths Production Unit, Learning & Teaching Solutions:
Becky Browne, Jim Campbell, Nicky Kempton, Bill Norman, Sharon Powell, Katie Sayce, Penny Tee

Alison Cadle	*TEX Consultant*
Michael Goldrei	*Cover Design Consultant*
Vicki McCulloch	*Cover Designer*

The external assessor was:

Jeff Paris	*Professor of Pure Mathematics, University of Manchester*

The Course Team would like to acknowledge their reliance on the previous work of Alan Slomson and of Alex Wilkie, Professor of Mathematical Logic, University of Oxford. We would also like to thank the family of Alonzo Church for generously providing the photo of him used in this unit.

This publication forms part of an Open University course. Details of this and other Open University courses can be obtained from the Student Registration and Enquiry Service, The Open University, PO Box 197, Milton Keynes, MK7 6BJ, United Kingdom: tel. +44 (0)870 300 6090, e-mail general-enquiries@open.ac.uk

Alternatively, you may visit the Open University website at http://www.open.ac.uk where you can learn more about the wide range of courses and packs offered at all levels by The Open University.

To purchase a selection of Open University course materials, visit http://www.ouw.co.uk, or contact Open University Worldwide, Michael Young Building, Walton Hall, Milton Keynes, MK7 6AA, United Kingdom, for a brochure: tel. +44 (0)1908 858793, fax +44 (0)1908 858787, e-mail ouw-customer-services@open.ac.uk

The Open University, Walton Hall, Milton Keynes, MK7 6AA.

First published 2004. Reprinted as new edition 2007, with corrections.

Edited, designed and typeset by The Open University, using the Open University TEX System.

Printed and bound in the United Kingdom by The Charlesworth Group, Wakefield.

ISBN 978 0 7492 2274 1

3.1

CONTENTS

Introduction   4

1 The Answer to Leibniz's Question   4
   1.1 Representability   5
   1.2 Algorithmically undecidable theories   8

2 Gödel's First Incompleteness Theorem   10
   2.1 Recursively axiomatizable theories   10
   2.2 Complete Arithmetic is not recursively
        axiomatizable   14

3 An Answer to Hilbert's Question   16
   3.1 Peano Arithmetic   17
   3.2 Gödel's Second Incompleteness Theorem   22

Biographical Sketches   26

Suggestions for Further Reading   30

Summary   33

Objectives   34

Solutions to the Problems   35

Index   40

INTRODUCTION

This final unit brings together all the ideas introduced in the course. These ideas constitute the technical machinery that enables us to prove some very important theorems which answer what we have called Leibniz's and Hilbert's Questions. These theorems, Gödel's Incompleteness Theorems, are among the most profound intellectual discoveries of the twentieth century. Thus you should not be surprised if you find this unit hard going in places. The proofs of almost all the key theorems are given and you should read them to be convinced of their validity. However, our emphasis will be on the *content* of the theorems rather than on the details of their proofs. We hope that, after reading this unit, you will have a good understanding of what these theorems tell us about the scope and limitations of mathematical logic and algorithmic methods.

Section 1 answers Leibniz's Question. Section 2 proves Gödel's First Incompleteness Theorem. Section 3 provides an answer to Hilbert's Question and proves Gödel's Second Incompleteness Theorem. The Biographical Sketches and Suggestions for Further Reading promised in earlier units appear at the end of this unit. There are no Additional Exercises for this unit.

1 THE ANSWER TO LEIBNIZ'S QUESTION

Recall that in *Unit 1* we posed Leibniz's Question in the following way:

> Is there an algorithm for deciding which statements of number theory are true?

In *Unit 4* we introduced a formal language and its standard interpretation $\mathcal{N}$, which enabled us to reformulate Leibniz's Question as:

> Is there an algorithm for deciding which formulas of our formal language are true under the standard interpretation $\mathcal{N}$?

In *Unit 6* we introduced Complete Arithmetic (CA) as the set of sentences of this language which are true in $\mathcal{N}$, and so were able to restate Leibniz's Question as:

> Is there an algorithm for deciding which sentences of our formal language are in the set CA?

In *Unit 1* we saw that the intuitive idea of an algorithm can be replaced by the idea of a URM computation, and in *Unit 3* we saw that the set of URM-computable functions is the same as the set of recursive functions. In order to reformulate Leibniz's Question in terms of recursive functions, which are functions of k-tuples of natural numbers, rather than in terms of formulas of our formal language, we saw in *Unit 4* how we can code these formulas using natural numbers called the Gödel numbers of the formulas. Then, making use of the idea of a recursive set rather than a recursive function, we were able in *Unit 4* to reformulate Leibniz's Question as:

Recall from *Unit 2* that a set of numbers A is said to be recursive if its characteristic function χ_A is recursive, where χ_A is defined by

$$\chi_A(n) = \begin{cases} 1, & \text{if } n \in A, \\ 0, & \text{if } n \notin A. \end{cases}$$

> Is the set of Gödel numbers of those formulas which are true in the standard interpretation $\mathcal{N}$ a recursive set?

Combining this with the definition of CA, we can now restate Leibniz's Question as:

> Is the set of Gödel numbers of the sentences in CA recursive?

It is this question that we can now answer in this section.

1.1 Representability

As we have said, Gödel numbers enable us to apply ideas about recursive sets to sets of formulas. However, as we saw in *Unit 7*, they have a much more subtle use. We can interpret formulas of our formal language as expressing properties of numbers; and, since we have assigned a unique Gödel number to each formula, we can therefore interpret formulas as expressing properties of formulas. This opens the possibility of finding formulas which refer to themselves, and Gödel's Diagonal Lemma shows that this possibility is realized for every theory which extends Q. As we shall shortly see, it is this which enables us, in Subsection 1.2, to give a negative answer to Leibniz's Question.

Theorem 3.5 of *Unit 7*.

To this end, we need some preliminary results about *representable sets*. We prove these in this subsection.

Definition 1.1

A set A is said to be *representable* in a theory T if its characteristic function is representable in T.

The Representability Theorem tells us that total recursive functions are representable in each theory T which extends Q. There is an equivalent result for recursive sets.

Theorem 2.5 of *Unit 7*.

Theorem 1.1

Let T be a theory which extends Q. Then every recursive set is representable in T.

Proof

If A is a recursive set, its characteristic function χ_A is recursive as well as total. Therefore, by the Representability Theorem, χ_A is representable in T, and so A is representable in T. ∎

If A is a recursive set, Theorem 1.1 tells us that there is a formula ϕ_A which represents its characteristic function χ_A in the theory T. The next result tells us that we can use such a formula ϕ_A to construct another formula which defines the elements of A in a more direct way.

Theorem 1.2

Let T be a theory which extends Q and let A be a set of natural numbers which is representable in T. Then there is a formula δ_A, in which x is the only variable which may occur freely, such that, for each natural number n:

(a) if $n \in A$ then $\vdash_T \delta_A(\mathbf{n})$;

(b) if $n \notin A$ then $\vdash_T \neg\delta_A(\mathbf{n})$.

Recall that **n** is the term consisting of the symbol **0** followed by n occurrences of $'$.

Proof

Suppose that ϕ_A is a formula which represents the characteristic function χ_A of A in the theory T. Hence, for all natural numbers n and k, if $\chi_A(n) = k$ then, by the definition of representability:

(i) $\vdash_T \phi_A(\mathbf{n}, \mathbf{k})$

(ii) $\vdash_T \forall y\, (\phi_A(\mathbf{n}, y) \rightarrow y = \mathbf{k})$

As χ_A is a characteristic function, the only significant values of k are 0 and 1.

We let δ_A be the formula $\phi_A(x, \mathbf{1})$. We show that this formula satisfies conditions (a) and (b) of the theorem.

Suppose, first, that $n \in A$. Then $\chi_A(n) = 1$ and hence, by (i), $\vdash_T \phi_A(\mathbf{n}, \mathbf{1})$, that is, $\vdash_T \delta_A(\mathbf{n})$. So condition (a) holds.

Suppose, second, that $n \notin A$. Then $\chi_A(n) = 0$ and hence, by (ii),

$$\vdash_T \forall y\, (\phi_A(\mathbf{n}, y) \rightarrow y = \mathbf{0})$$

Since T extends Q, it follows from Theorem 2.1 of *Unit 7* that

$$\vdash_T \neg \mathbf{1} = \mathbf{0}.$$

Now consider the following formal proof.

1	(1)	$\forall y\, (\phi_A(\mathbf{n}, y) \rightarrow y = \mathbf{0})$	Ass
2	(2)	$\neg \mathbf{1} = \mathbf{0}$	Ass
3	(3)	$\phi_A(\mathbf{n}, \mathbf{1})$	Ass
1	(4)	$(\phi_A(\mathbf{n}, \mathbf{1}) \rightarrow \mathbf{1} = \mathbf{0})$	UE, 1
1,3	(5)	$\mathbf{1} = \mathbf{0}$	Taut, 3, 4
1,2,3	(6)	$(\mathbf{1} = \mathbf{0}\ \&\ \neg \mathbf{1} = \mathbf{0})$	Taut, 2, 5
1,2	(7)	$(\phi_A(\mathbf{n}, \mathbf{1}) \rightarrow (\mathbf{1} = \mathbf{0}\ \&\ \neg \mathbf{1} = \mathbf{0}))$	CP, 6
1,2	(8)	$\neg \phi_A(\mathbf{n}, \mathbf{1})$	Taut, 7

It follows that

$$\forall y\, (\phi_A(\mathbf{n}, y) \rightarrow y = \mathbf{0}), \neg \mathbf{1} = \mathbf{0} \vdash \neg \phi_A(\mathbf{n}, \mathbf{1})$$

Hence, as the formulas $\forall y\, (\phi_A(\mathbf{n}, y) \rightarrow y = \mathbf{0})$ and $\neg \mathbf{1} = \mathbf{0}$ are derivable in T, so also is the formula $\neg \phi_A(\mathbf{n}, \mathbf{1})$. That is $\vdash_T \neg \delta_A(\mathbf{n})$. So condition (b) also holds. ∎

It turns out that the converse of this theorem also holds. We do not use the converse in this unit, but if you want practice in finding formal proofs, you may wish to tackle the following problem which asks you to establish it.

Problem 1.1 _____

Let T be a theory which extends Q and let A be a set of natural numbers such that there is a formula δ_A, with x as its only free variable, for which conditions (a) and (b) of Theorem 1.2 hold. Deduce that the set A is representable in T. *Hint:* Show that, given the assumptions of this problem, the characteristic function of the set A is represented by the formula $((\delta_A(x)\ \&\ y = \mathbf{1}) \vee (\neg \delta_A(x)\ \&\ y = \mathbf{0}))$.

_____

Theorem 1.2 together with the result of Problem 1.1 show that, for a theory T which extends Q, the following conditions on a set A of natural numbers are equivalent.

(a) The set A is representable in T.

(b) There is a formula δ_A whose only free variable is x such that, for each natural number n, if $n \in A$ then $\vdash_T \delta_A(\mathbf{n})$ and if $n \notin A$ then $\vdash_T \neg \delta_A(\mathbf{n})$.

Recall that, in *Unit 6*, we defined a theory T to be *consistent* if there is no sentence Φ such that both $\vdash_T \Phi$ and $\vdash_T \neg \Phi$. We saw that if a theory is *not* consistent then every sentence is derivable in that theory. So an inconsistent theory is of no interest. Thus in subsequent results we build in the assumption that we are dealing with a consistent theory.

Problem 3.2 of *Unit 6*.

We are now ready to prove a key result. It is our first application of Gödel's Diagonal Lemma. We shall see that it has many important consequences.

Theorem 1.3

Let T be a consistent theory which extends Q. Then the set $GN(T)$ of Gödel numbers of the theorems of T is not representable in T.

Proof

Let T be a consistent theory which extends Q. We are aiming to show that there is no formula which represents the set $GN(T)$ of Gödel numbers of the theorems of T within T. We show that no such formula exists by demonstrating that the assumption that such a formula exists leads to a contradiction.

So let us assume that $GN(T)$ is representable in T. Then, by Theorem 1.2, there is a formula θ_T whose only free variable is x such that, for each natural number n,

$$\left.\begin{array}{l} n \in GN(T) \implies \vdash_T \theta_T(\mathbf{n}), \\ n \notin GN(T) \implies \vdash_T \neg\theta_T(\mathbf{n}). \end{array}\right\} \tag{1.1}$$

Now, as $GN(T)$ is the set of the Gödel numbers of the theorems of T, we have that, for each formula ϕ,

$$\left.\begin{array}{l} \vdash_T \phi \implies \gamma(\phi) \in GN(T), \\ \text{not } \vdash_T \phi \implies \gamma(\phi) \notin GN(T). \end{array}\right\} \tag{1.2}$$

By (1.1) and (1.2), we have

$$\left.\begin{array}{l} \vdash_T \phi \implies \vdash_T \theta_T(\ulcorner\phi\urcorner), \\ \text{not } \vdash_T \phi \implies \vdash_T \neg\theta_T(\ulcorner\phi\urcorner). \end{array}\right\} \tag{1.3}$$

Recall that $\gamma(\phi)$ is the Gödel number of ϕ and that $\ulcorner\phi\urcorner$ is the term consisting of of the symbol $\mathbf{0}$ followed by $\gamma(\phi)$ occurrences of $'$. In the standard interpretation $\mathcal{N}$ this term is taken as referring to the number $\gamma(\phi)$.

We now apply Gödel's Diagonal Lemma to the formula $\neg\theta_T$. This tells us that there is a sentence $G_{\neg\theta_T}$ such that

$$\vdash_T (G_{\neg\theta_T} \leftrightarrow \neg\theta_T(\ulcorner G_{\neg\theta_T}\urcorner)) \tag{1.4}$$

To simplify the notation, we shall write G instead of $G_{\neg\theta_T}$, so that (1.4) can be rewritten as

$$\vdash_T (G \leftrightarrow \neg\theta_T(\ulcorner G\urcorner)) \tag{1.5}$$

We now ask whether the sentence G is a theorem of T.

Suppose first that

$$\vdash_T G \tag{1.6}$$

Then, by (1.3),

$$\vdash_T \theta_T(\ulcorner G\urcorner) \tag{1.7}$$

But then, by an application of the Tautology Rule to (1.5) and (1.7), we deduce that

$$\vdash_T \neg G \tag{1.8}$$

Now (1.6) and (1.8) contradict our assumption that T is consistent. Thus supposition (1.6) is wrong, and so we have shown that

$$\text{not } \vdash_T G \tag{1.9}$$

However, now it follows from (1.3) that

$$\vdash_T \neg\theta_T(\ulcorner G\urcorner) \tag{1.10}$$

and, by an application of the Tautology Rule to (1.5) and (1.10), we deduce that

$$\vdash_T G \tag{1.11}$$

But this contradicts (1.9). We have been led to this contradiction because we made the assumption that $GN(T)$ is representable in T. This assumption must therefore be wrong. Thus we have proved that $GN(T)$ is not representable in T. ∎

The power of this theorem will become evident from the consequences that we draw from it in the following pages.

Problem 1.2 _____

In the proof of Theorem 1.3 we made two applications of the Tautology Rule. Specify the two tautologies which justify these applications, and check that they are tautologies.

_____

1.2 Algorithmically undecidable theories

We are now able to give an answer to Leibniz's Question. Recall that in the introduction to this section we reformulated this question as:

Is the set of Gödel numbers of the sentences in CA recursive?

Complete Arithmetic, CA, is a consistent theory which extends Q. Hence, by Theorem 1.1, every recursive set is representable in CA and, by Theorem 1.3, the set $GN(\text{CA})$ is not representable in CA. We can immediately deduce that $GN(\text{CA})$ is not a recursive set. If a sentence is a theorem of CA, it is true in the standard interpretation $\mathcal{N}$ and hence is in CA. Thus $GN(\text{CA})$ is in fact the set of Gödel numbers of all the sentences in CA. Thus we have obtained a negative answer to Leibniz's Question. For the record we formulate this answer as a theorem.

_____

Theorem 1.4 *Negative Answer to Leibniz's Question*

The set of Gödel numbers of the sentences which are true in the standard interpretation $\mathcal{N}$ is not recursive.

_____

The great interest and importance of this result springs from our work in *Units 1* to *3*, where we showed the connection between recursive functions and computable functions and where we gave evidence for Church's Thesis, which identifies the notions of an algorithmically computable function and of a recursive function. Note also that sentences which we have described as being true 'in the standard interpretation' would more commonly simply be described as being true. Thus we are able to restate Theorem 1.4 in a perhaps more dramatic way which makes clearer its relationship to Leibniz's Question.

_____

Theorem 1.4 *Negative Answer to Leibniz's Question (Restated)*

There is no algorithm for deciding which statements of number theory are true.

_____

In one sense, this answer to Leibniz's Question is disappointing. If an algorithm for deciding which statements of number theory are true existed, this would enable us to settle all the outstanding questions of number theory in a routine way. On the other hand, it is comforting to know that human ingenuity will always be needed, and that mathematicians cannot be replaced by computers!

Although it may not come as a surprise that there is no algorithm for deciding which sentences are theorems of Complete Arithmetic, it is important to note that the argument used to derive Theorem 1.4 applies also to much weaker theories. Our proof of Theorem 1.4 relies only on Theorems 1.1 and 1.3. Thus it can be generalized immediately to cover any consistent theory which extends Q.

Theorem 1.5

Let T be a consistent theory which extends Q. Then the set of Gödel numbers of the sentences which are theorems of T is not recursive.

In particular, this result applies to the theory Q itself.

Theorem 1.6

There is no algorithm for deciding which sentences are theorems of Q.

This is, in some ways, a more surprising result than Theorem 1.4 because, as we have seen, Q is such a weak theory. We discovered in *Unit 6* that from the axioms of Q we cannot derive, for example, the sentence $\forall x \forall y (x + y) = (y + x)$ which expresses the fact that addition of natural numbers is commutative. We gave several more examples of simple facts that cannot be derived from the axioms of Q, showing that it is a weak theory. None the less, Theorem 1.6 shows that the set of theorems of Q is sufficiently complicated that it (or, more strictly, the set of the Gödel numbers of the theorems of Q) is a non-recursive set.

It is even more significant that Q has only a finite number of axioms, indeed just seven. This enables us to deduce from Theorem 1.6 a significant result about quantifier logic, which we anticipated at the end of Subsection 1.2 of *Unit 5*. This result is Church's Theorem, which we shall prove shortly, after proving a couple of preliminary results.

Theorem 1.7

There is no algorithm for deciding, in general, given formulas $\phi_1, \phi_2, \ldots, \phi_k$ and ψ, whether $\phi_1, \phi_2, \ldots, \phi_k \vdash \psi$.

Proof

Let $\sigma_1, \sigma_2, \ldots, \sigma_7$ be the seven sentences which make up the axioms of Q. By Theorem 1.6 there is no algorithm for deciding, given a sentence σ, whether $\vdash_Q \sigma$, that is whether $\sigma_1, \sigma_2, \ldots, \sigma_7 \vdash \sigma$. Hence there cannot be an algorithm for deciding, in general, whether $\phi_1, \phi_2, \ldots, \phi_k \vdash \psi$. ∎

Theorem 1.8

There is no algorithm for deciding, in general, given a formula ψ, whether $\vdash \psi$.

Proof

Since $\phi_1, \phi_2, \ldots, \phi_k \vdash \psi$ if and only if $\vdash ((\cdots (\phi_1 \ \& \ \phi_2) \ \& \ \cdots \ \& \ \phi_k) \to \psi)$, an algorithm for deciding whether a given formula can be derived from no assumptions would enable us to decide, given formulas $\phi_1, \phi_2, \ldots, \phi_k$ and ψ, whether $\phi_1, \phi_2, \ldots, \phi_k \vdash \psi$, contradicting Theorem 1.7. ∎

Now sentences which are derivable from no assumptions are, by the Correctness and Adequacy Theorems of *Unit 6*, precisely those sentences which are logically valid, that is, true in all possible interpretations. Thus, from Theorem 1.8, we can deduce the following result.

Theorem 1.9 Church's Theorem

There is no algorithm for deciding which sentences of quantifier logic are logically valid.

This theorem was first proved by Alonzo Church in 1936. It is often paraphrased as 'quantifier logic is algorithmically undecidable'.

We explained in *Unit 5*, at the end of Subsection 1.2, why we could not simply use a 'logical consequence' rule for quantifier logic analogous to the Tautology Rule for propositional logic. We have now proved the theorems which justify our remarks that such a 'logical consequence' rule would fail to have the algorithmic character that we require for a formal system.

2 GÖDEL'S FIRST INCOMPLETENESS THEOREM

In this section we shall look at axioms for number theory within our formal system. In Subsection 3.2 of *Unit 6* we gave some criteria for a set S of axioms for number theory. These included the following.

- There should be an algorithm to determine whether a given sentence is in S.

When we first talked about mathematical proof in *Unit 5* we said that one requirement of a formal system — the machine-checkability requirement — is that there should be an algorithm to decide whether a sequence of formulas satisfies the requirement of being a formal proof. The rules of proof that we introduced in *Units 5* and *6* all meet this requirement. But if we want there to be an algorithm to decide whether a sequence of formulas is a formal proof *from a given set of axioms*, then we also require that there should be an algorithm to decide whether or not a formula is an axiom; hence the above criterion.

The other criteria were that all the axioms in S should be true in the standard interpretation $\mathcal{N}$, that we should be able to derive as many theorems as possible from the axioms and that the set of axioms should be as simple as possible.

The results of the previous section, in particular Theorem 1.4, tell us that the set CA, the set of all sentences of the formal language true in the standard interpretation $\mathcal{N}$, does not meet this criterion. However, this still leaves open the question of whether there is some set S of axioms for CA which does meet the criteria. This is the issue which we look at in this section, culminating in Gödel's First Incompleteness Theorem which resolves it. The criterion that there should be an algorithm to determine whether a given sentence is in the set of axioms is investigated in Subsection 2.1 and the application of this idea to number theory is discussed in Subsection 2.2.

2.1 Recursively axiomatizable theories

In this subsection we look at theories with a set S of axioms for which there is an algorithm to determine whether a given sentence is in S. We make this more precise with the following definition.

We shall often refer to a set of axioms whose Gödel numbers form a recursive set as a *recursive set of axioms*. Thus we can say that a theory is recursively axiomatizable if it has a recursive set of axioms. We shall also refer to a set of theorems whose Gödel numbers form a recursive set as a *recursive set of theorems*.

In framing Definition 2.1 we are making a subtle distinction between a theory which has a recursive set of *axioms* and a theory which has a recursive set of *theorems*. The example given by the theory Q shows us that these are not the same thing. The set of axioms of Q is finite and hence is recursive, but Theorem 1.6 tells us that the set of theorems of Q is not recursive. Another example of this kind is given by Theorem 1.8. Thus, despite the negative answer to Leibniz's Question given by Theorem 1.4, there remains the possibility that CA has a recursive set of axioms.

To investigate whether or not such a recursive set of axioms for CA exists, we begin by looking generally in this subsection at properties of theories with a recursive set of axioms or, more precisely, at the corresponding sets of Gödel numbers of theorems. First we introduce a new way of describing sets of natural numbers.

> Definition 2.2 Recursive enumerability
>
> A set X of natural numbers is said to be *recursively enumerable* if either X is empty or X can be expressed in the form
>
> $$X = \{f(0), f(1), f(2), \ldots\}$$
>
> where f is a total recursive function.

In other words, a non-empty set of natural numbers is recursively enumerable if it is the set of values taken by a total recursive function. Note that our notation is not intended to imply that the values $f(0), f(1), f(2), \ldots$ are all distinct. Thus, as shown by Example 2.1 below, a recursively enumerable set can be finite, while Example 2.3 provides an instance of an infinite set enumerated by a recursive function whose values are not all distinct. It is convenient to regard the empty set as being recursively enumerable, and, as this set cannot be the set of values taken by a total function, in our definition we need to specify separately that the empty set is recursively enumerable.

By Church's Thesis, a function f is recursive if and only if there is an algorithm for computing its values. So a set is recursively enumerable if and only if either it is empty or there is an algorithm for enumerating (or listing) its members.

There is an important, but quite subtle, difference between a recursive set and a recursively enumerable set, as we shall see later.

Example 2.1

The set $\{0\}$ is recursively enumerable. The zero function $\text{zero} : n \longmapsto 0$ is recursive and the set of values taken by this function is just the set $\{0, 0, 0, \ldots\} = \{0\}$. ◆

Example 2.2

The set $\{0, 1, 4, 9, \ldots\}$ of squares is recursively enumerable. The function $f : n \longmapsto n^2$ is recursive and the set of squares is the set of values taken by this function. ◆

Example 2.3

The set Y of all natural numbers whose decimal representations occur as consecutive digits somewhere to the right of the decimal point in the infinite decimal expansion of π is a recursively enumerable set.

We demonstrate this by means of an informal argument. As

$$\pi = 3.14159\ldots$$

the set Y includes, for example, 5 (coming from the underlined part of $3.14\underline{5}9\ldots$), 415 (coming from $3.1\underline{415}9\ldots$) and so on.

There are well-known algorithms, based for example on convergent series, that enable us to compute the decimal expansion of π to as many places as we wish. We can use such an algorithm to enumerate all finite strings of digits which occur after the decimal point as follows:

Step 1 Generate the first digit after the decimal point: $\pi = 3.1\ldots$ so the list begins $\{1, \ldots\}$.

Step 2 Generate the next digit: $\pi = 3.14\ldots$ so we add to the list 4 and 14 giving $\{1, 4, 14, \ldots\}$.

Step 3 Generate the next digit: $\pi = 3.141\ldots$ so we add to the list $1, 41$ and 141 giving $\{1, 4, 14, 41, 141, \ldots\}$.

$\vdots$

It doesn't matter that the string 1 gets repeated by this process, here at *Step 3*. The values of the enumerating function need not be distinct.

Given an algorithm for generating the decimal expansion of π as far as we like, we have an algorithm which enables us to continue the list as far as we like. Thus if f is defined by $f(n) =$ the nth number in the list (counting 1 as $f(0)$), we see that Y is the set of values taken by a total computable function f and hence Y is recursively enumerable. ◆

Our argument in Example 2.3 is an example of the use of Church's Thesis. We have relied on the fact that we can describe informally an algorithm for computing the values of f to deduce that f is a recursive function. In this way we have avoided the technical complexities of a more precise proof that f is a recursive function. In what follows we shall continue to use informal arguments of this kind.

Problem 2.1 _____

Show that the set $\{2, 3, 8\}$ is recursively enumerable.

Problem 2.2 _____

Show that every finite set of natural numbers is recursively enumerable.

Problem 2.3 _____

Show that every recursive set of natural numbers is recursively enumerable. *Hint:* If A is a recursive set of natural numbers, its characteristic function χ_A is a total recursive function. Use χ_A to devise a total recursive function whose set of values is precisely the set A.

From our point of view the most important examples of recursively enumerable sets are those given by the next theorem. Before coming to this, we need to remind ourselves that in *Unit 4*, Subsection 2.2, we showed how to assign Gödel numbers to formulas, and to indicate how we can extend this numbering to formal proofs.

Recall that a formal proof consists of a numbered finite sequence of formulas together with justifications explaining how each formula in the sequence has been obtained using the rules of proof. We already know how to assign a Gödel number $\gamma(\phi)$ to each formula ϕ. This Gödel numbering exploits the fact that the formulas consist of finite sequences of the basic symbols, and that these basic symbols are drawn from a finite list. The justifications, such as Taut, $3, 5$ and Ass, that occur in a formal proof also consist of finite

Note that we don't have to concern ourselves with the assumption numbers, given to the left of each line of a proof, as these can be calculated from the justifications.

sequences of symbols drawn from the finite list made up of the letters of the English alphabet, the standard numerals $0, 1, \ldots, 9$ and a comma. Thus, by assigning Gödel numbers to each of these symbols, we can duplicate the technique for assigning Gödel number to formulas to enable us to assign Gödel numbers to justifications. Let us denote the Gödel number assigned to a justification j as $\delta(j)$. Then, for any formal proof F given by

(1) ϕ_1 j_1

(2) ϕ_2 j_2

$\vdots$

(k) ϕ_k j_k

we assign to it the Gödel number

$$\Gamma(F) = 2^{\gamma(\phi_1)} 3^{\delta(j_1)} \ldots p_{2k-1}^{\gamma(\phi_k)} p_{2k}^{\delta(j_k)},$$

where p_i is the ith prime number.

It should now be evident that, if we are given a natural number n, we can first compute whether or not n is the Gödel number of a formal proof, and if it is we can compute from n the Gödel numbers of the formulas which are the assumptions in force on the last line of the proof.

We are now ready to prove the key theorem of this subsection.

Theorem 2.1

The set $GN(T)$ of Gödel numbers of the theorems of a recursively axiomatizable theory T form a recursively enumerable set.

Proof

Let T be a theory with a recursive set S of axioms.

The set of theorems of T is certainly not empty. For example, it includes the sentence

$\forall x_0\, x_0 = x_0$

We let n_0 be the Gödel number of this sentence.

We define a recursive function f by giving an informal description of an algorithm for computing the values of f.

Given $n \in \mathbb{N}$, first compute whether n is the Gödel number of a formal proof.

If not we put $f(n) = n_0$.

If n is the Gödel number of a formal proof, we can compute from n the Gödel numbers of the assumptions on which the last line of this proof depends. Then, as S is a recursive set of axioms, we can compute whether or not all these assumptions are in S. If all these assumptions are in S, then we put

$f(n) =$ the Gödel number of the formula on the last line of the proof;

if not, we put $f(n) = n_0$, as before.

Thus, for all $n \in \mathbb{N}$, $f(n)$ is the Gödel number of a formula which can be derived from assumptions which are in S, and hence $f(n)$ is the Gödel number of a theorem of T.

Conversely, if ϕ is a theorem of T, there is a formal proof, say F, of ϕ from assumptions in S. So $\Gamma(F)$ is the Gödel number of such a proof, and hence $f(\Gamma(F)) = \gamma(\phi)$, that is, $f(\Gamma(F))$ is the Gödel number of ϕ.

Thus $\{f(0), f(1), f(2), \ldots\}$ is the set $GN(T)$ of all the Gödel numbers of the theorems of T. Since f is a total recursive function, it follows that this set is recursively enumerable. ∎

Γ is Greek letter 'capital gamma', corresponding to the lower-case γ. We use the capital letter Γ to distinguish the Gödel number of a proof from the Gödel number $\gamma(\phi)$ of a formula ϕ.

An explicit expression for n_0 is given in Subsection 2.2 of *Unit 4*.

In giving the proof of Theorem 2.1, we have relied heavily on Church's Thesis. However, note that we have only used it in what we described in *Unit 3* as the practical sense, in which it expresses our confidence that we could turn the informal algorithm for computing the values of f, as described in the proof, into a precise proof that f is a recursive function.

2.2 Complete Arithmetic is not recursively axiomatizable

It follows from Theorem 2.1 that the set $GN(Q)$ of Gödel numbers of the theorems of Q is recursively enumerable. This is in contrast with the result of Theorem 1.6, which tells us that this set is not recursive. Thus $GN(Q)$ provides us with an example of a set which is recursively enumerable but not recursive. This means that, although we can enumerate this set as

$$GN(Q) = \{f(0), f(1), f(2), \ldots\}$$

where f is a recursive function, there is no algorithm for determining whether or not a given number is in this set. The moral to be drawn here is that the existence of an algorithm for *enumerating* the theorems of a recursively axiomatizable theory does not imply the existence of an algorithm for *deciding* whether or not a given sentence is a theorem: 'wait and see if it appears in the list' is not an algorithm for giving an answer to this question. If a sentence is a theorem then it will appear at some stage of the enumeration, and when it does appear we shall know that it is a theorem; but if a sentence is not a theorem, there may never be a stage when we can be sure of this, since the fact that a sentence has not so far appeared in the enumeration may not give us any information about whether it will appear later.

There is, however, one special case where we can deduce that a recursively axiomatizable theory has a recursive set of theorems. Recall that a theory T is said to be *complete* if, for every sentence Φ of the formal language, either $\vdash_T \Phi$ or $\vdash_T \neg\Phi$. We noted in Section 3 of *Unit 6* that if T is the set of all sentences which are true in some particular interpretation of the formal language then T is complete. In particular Complete Arithmetic, which is the set of all sentences true in the standard interpretation, is a complete theory.

Definition 3.4 of *Unit 6*.

Theorem 2.2

A complete recursively axiomatizable theory has a recursive set of theorems.

14

Proof

Let T be a complete recursively axiomatizable theory.

Problem 3.2 of *Unit 6*.

We note first that, if T is not consistent, then every sentence is a theorem of T and so, in a very uninteresting way, there is an algorithm for deciding which sentences are theorems of T. Thus from now on we need only consider the case where T is a consistent theory.

By Theorem 2.1, there is an algorithm for enumerating the theorems of T. Suppose that this algorithm produces the list

$$\Phi_0, \Phi_1, \Phi_2, \ldots$$

of all the theorems of T. We describe an algorithm for deciding whether or not a given sentence Φ is a theorem of T.

Given a sentence Φ, we know that, as T is complete, either Φ or $\neg\Phi$ is a theorem of T, and that, as T is consistent, they cannot both be theorems of T. Thus after a finite number of steps of the enumeration, one of Φ and $\neg\Phi$ must appear in the list. If it is Φ which appears, then we can deduce that Φ is a theorem of T. If it is $\neg\Phi$ which appears, we can deduce that Φ does not occur anywhere in the list and so is not a theorem of T. So in either case, after a finite number of steps the algorithm terminates with a correct answer as to whether or not Φ is a theorem of T. ∎

We now come to the main theorem of the course.

Theorem 2.3 Gödel's First Incompleteness Theorem

There is no complete and consistent recursively axiomatizable extension of Q.

Proof

By Theorem 2.2, if T were a complete and consistent recursively axiomatizable theory which extends Q, then T would have a recursive set of theorems. But this contradicts Theorem 1.5. Hence no such complete and consistent extension of Q exists. ∎

Theorem 2.4

Complete Arithmetic is not recursively axiomatizable.

Proof

Since Complete Arithmetic is a complete and consistent extension of Q, it is an immediate consequence of Theorem 2.3 that Complete Arithmetic is not recursively axiomatizable. ∎

Theorem 2.3 is known as an 'incompleteness' theorem because it may be reformulated as follows.

> If T is a consistent recursively axiomatizable extension of Q, then T is not complete.

Gödel first announced this theorem publicly at a meeting in Königsberg in September 1930. The details of the proof were given in his paper which was published in the following year. Although the proof we have given here uses essentially the same methods and ideas as Gödel's first published proof, it differs from it in some significant respects.

For bibliographical details, see the Suggestions for Further Reading at the end of this unit.

First, in his 1931 paper Gödel points out that his proof works for any theory sufficiently powerful that every recursive relation is representable in it. However, in 1931 it was not realized that there are systems as weak as the finitely axiomatizable theory Q in which all total recursive functions are representable. Thus Gödel took as his base theory one which is significantly more powerful than Q.

We shall discuss this theory in Section 3.

Second, our proof of Theorem 2.3 depends on deducing it from Theorem 1.5, and hence ultimately from Theorem 1.3. It is in our proof of Theorem 1.3 that we made heavy use of Gödel's ideas and in particular the device of Gödel numbering and Gödel's Diagonal Lemma. In contrast, Gödel gave a direct proof of Theorem 2.3 in which he uses his methods to show how to construct, for a given recursively axiomatized consistent extension T of Q, a sentence σ_T such that neither σ_T nor $\neg\sigma_T$ is a theorem of T. This shows directly that T is not complete. Moreover, it can be seen from the construction of the sentence σ_T that $\neg\sigma_T$ is true in the standard interpretation. Thus Gödel's proof provides an example of a sentence which is true (in the standard interpretation) but which is not a theorem of T.

We say more about this in the next section.

What was missing in 1931 was an appreciation of Theorem 2.2, which for us enables us to deduce Theorem 2.3 very readily from Theorem 1.5. From our viewpoint, with our experience of recursive and recursively enumerable sets, it is rather surprising that the truth of Theorem 2.2, whose proof to us seems very straightforward, was overlooked at the time. Gödel's First Incompleteness Theorem shows that Complete Arithmetic cannot be recursively axiomatized. This result came as a considerable surprise. However, Theorem 2.2 shows that if Complete Arithmetic could be recursively axiomatized, then there would be an algorithm for deciding which sentences of number theory are true in the standard interpretation, a result that even in 1931 might have seemed implausible.

3 AN ANSWER TO HILBERT'S QUESTION

In this section we shall answer Hilbert's Question:

> Can the consistency of number theory be proved using only non-dubious principles of finitary reasoning?

One of our purposes in setting up our formal system in *Units 4* to *6* was to provide a framework for proof based on finitary reasoning. Within this framework, we need to choose a set of axioms for number theory so that we can investigate Hilbert's Question with the corresponding theory. To ensure finitary reasoning, we want a theory with a set of axioms which is at the very least recursive, so that Complete Arithmetic is unsuitable. At the other end of the spectrum, the theory Q is plainly too weak to be considered as number theory for this purpose. In Subsection 3.1 we shall look at the theory *Peano Arithmetic* which provides a useful and important compromise between these unsuitable theories and in Subsection 3.2 we answer Hilbert's Question for this theory.

3.1 Peano Arithmetic

Gödel's First Incompleteness Theorem tells us that no consistent recursively axiomatizable extension of Q is complete. This theorem applies to the theory Q itself. But, of course, it is no surprise that Q is not complete. In Section 3 of *Unit 6* we noted many examples of sentences which are true in the standard interpretation $\mathcal{N}$ but are not theorems of Q. These include such simple sentences as $\forall x \,\neg\, x' = x$ and $\forall x \, \forall y \,(x + y) = (y + x)$. Thus Q is a weak theory. In this subsection we introduce a much more powerful consistent recursively axiomatizable theory whose theorems include many more sentences which are true in the standard interpretation.

From your studies in the Number Theory part of M381, you will probably find it easy to see what is missing from Q that makes it such a weak theory. One of the most common methods of proof in number theory is Mathematical Induction, but there is nothing in the theory Q which corresponds to this.

The Principle of Mathematical Induction is formulated in *Unit 1* of Number Theory in the following way.

The surprising thing about Q is that, despite being a weak theory, its theorems do not form a recursive set.

Principle of Mathematical Induction

Let $P(n)$ be a proposition depending on a natural number n. If:

(a) $P(0)$ is true;

(b) for each natural number k, if $P(k)$ is true then $P(k + 1)$ is true;

then $P(n)$ is true for all $n \in \mathbb{N}$.

We have changed the wording slightly from that given in *Unit 1* of Number Theory to allow for the fact that we are dealing with the set of natural numbers, starting at 0, rather than with the positive integers, starting at 1.

The theory we are about to study attempts to incorporate this principle into its axioms. What we need is a formal counterpart to the notion of a 'proposition depending on a natural number n' used to formulate the principle above.

We have seen in *Units 4* and *5* that, if ϕ is a formula in which the variable x occurs freely, then whether or not ϕ is true in the standard interpretation $\mathcal{N}$ depends on how we interpret x, that is, on which natural number we regard x as representing. In this way we can think of ϕ as expressing a proposition which depends on a natural number.

The formula ϕ will be true in the standard interpretation when x is interpreted as the natural number n if and only if $\phi(\mathbf{n}/x)$ is true in this interpretation, where $\phi(\mathbf{n}/x)$ denotes the formula obtained by substituting the term $\mathbf{n}$ for all free occurrences of the variable x in ϕ. To fit in better with the notation used in expressing the Principle of Mathematical Induction, we shall write $\phi(\mathbf{n})$ instead of $\phi(\mathbf{n}/x)$. More generally, we shall write $\phi(\tau)$ instead of $\phi(\tau/x)$. It follows that $\phi(x)$ is the same as the formula ϕ; writing it as $\phi(x)$ emphasizes that we are interested in the fact that the variable x has free occurrences in ϕ, so that ϕ can be interpreted as expressing a property of natural numbers.

The notation $\phi(\mathbf{n}/x)$, or more generally $\phi(\tau/x)$, was introduced in *Unit 5*.

The notation $\phi(\mathbf{n})$ was introduced (albeit in a different context) in *Unit 7*.

Thus we are now able to give a formal counterpart of the Principle of Mathematical Induction, as follows.

Let ϕ be a formula in which the variable x may occur freely. Then the formulas $\phi(\mathbf{0})$, $\forall x \,(\phi(x) \rightarrow \phi(x'))$ imply $\forall x \, \phi(x)$.

Here, $\phi(\mathbf{0})$ is the counterpart of '$P(0)$ is true', $\forall x \,(\phi(x) \rightarrow \phi(x'))$ corresponds to 'for each natural number k, if $P(k)$ is true then $P(k + 1)$ is true' and $\forall x \, \phi(x)$ corresponds to the conclusion '$P(n)$ is true for all $n \in \mathbb{N}$'.

Thus one way to incorporate the Principle of Mathematical Induction into our theory would be to take sentences of the form

$$((\phi(\mathbf{0}) \mathbin{\&} \forall x\,(\phi(x) \to \phi(x'))) \to \forall x\,\phi(x))$$

as axioms. In fact, in practice we shall also wish to use such axioms in the cases where the formula ϕ has, in addition, free occurrences of variables other than x. In such cases we need to prefix universal quantifiers for these other variables in order to express the truth of the above formula however these other variables are interpreted. For example, if ϕ is a formula in which x and y, but no other variables, may occur freely, then we need to take the sentence

$$\forall y\,((\phi(\mathbf{0}, y) \mathbin{\&} \forall x\,(\phi(x, y) \to \phi(x', y))) \to \forall x\,\phi(x, y))$$

as an axiom. The theory we get by adding to the axioms of Q all sentences of this form is called *Peano Arithmetic* and is denoted by PA.

Definition 3.1 Peano Arithmetic

Peano Arithmetic is the theory PA whose axioms consist of the axioms $Q1$ to $Q7$ of Q together with all sentences of the formal language obtained from formulas of the form

$$((\phi(\mathbf{0}) \mathbin{\&} \forall x\,(\phi(x) \to \phi(x'))) \to \forall x\,\phi(x))$$

by prefixing universal quantifiers for all variables which occur freely. Axioms of this latter form are called *induction axioms*.

Giuseppe Peano 1858–1932.

The German mathematician Richard Dedekind (1831–1916) was first to characterize the natural numbers by axioms which included the Principle of Mathematical Induction. The Italian mathematician Giuseppe Peano devised a more formal set of axioms for number theory in 1889 and the theory PA is named after him.

Like the axioms of Q, the induction axioms are all true in the standard interpretation $\mathcal{N}$. Hence all the theorems of PA are true in the standard interpretation and hence PA is a consistent theory. It is evidently an extension of Q. Although there are infinitely many induction axioms, it is not difficult to see that there is an algorithm for determining whether a formula has the form of an induction axiom. Hence the axioms of PA form a recursive set. Thus PA is a consistent recursively axiomatizable extension of Q, and hence, by Theorem 2.3, PA is not complete.

Although incomplete, PA is a very much more powerful theory than Q, because it includes all the induction axioms. The proofs of a great number of the known theorems of number theory can be given in the theory PA. We do not have the time to go very far in exploring the strength of this system. The evidence that had accumulated by 1930 certainly suggested that it might be possible to derive all the true sentences of number theory, that is all the sentences of our formal language which are true in the standard interpretation, in the theory PA. That is why it was a shock when Gödel published the proof of his First Incompleteness Theorem in 1931. Gödel's paper of 1931 was based on a formal system which is essentially the same as PA.

Example 3.1

Our first example of a formal proof using an induction axiom shows that, with its aid, we can derive the axiom $Q3$ from the remaining axioms of PA.

We let $\phi(x)$ be the formula $(\neg x = \mathbf{0} \to \exists y \, x = y')$. Thus $\phi(\mathbf{0})$ is the formula

$$(\neg \mathbf{0} = \mathbf{0} \to \exists y \, \mathbf{0} = y') \tag{3.1}$$

and $(\phi(x) \to \phi(x'))$ is the formula

$$((\neg x = \mathbf{0} \to \exists y \, x = y') \to (\neg x' = \mathbf{0} \to \exists y \, x' = y')) \tag{3.2}$$

We show that (3.1) and (3.2) are both theorems of PA, and then, by using the appropriate induction axiom, that $\forall x \, \phi(x)$ is a theorem of PA.

	(1)	$\mathbf{0} = \mathbf{0}$	II
	(2)	$\phi(\mathbf{0})$	Taut, 1
	(3)	$x' = x'$	II
	(4)	$\exists y \, x' = y'$	EI, 3
	(5)	$(\phi(x) \to \phi(x'))$	Taut, 4
	(6)	$\forall x \, (\phi(x) \to \phi(x'))$	UI, 5
7	(7)	$((\phi(\mathbf{0}) \,\&\, \forall x \, (\phi(x) \to \phi(x'))) \to \forall x \, \phi(x))$	Ass
7	(8)	$\forall x \, \phi(x)$	Taut, 2, 6, 7

Note that the assumption on line 7 is an induction axiom. Hence $\vdash_{\mathrm{PA}} \forall x \, \phi(x)$.

Now let PA' be the theory obtained from PA by removing the axiom $Q3$. The formal proof above shows that

$$\vdash_{\mathrm{PA}'} \forall x \, (\neg x = \mathbf{0} \to \exists y \, x = y')$$

That is, axiom $Q3$ is a theorem of PA'. Thus the inclusion of $Q3$ in the set of axioms of PA was not necessary. ♦

Example 3.2

We show that

$$\vdash_{\mathrm{PA}} \forall x \, (\mathbf{0} + x) = x$$

Let $\phi(x)$ be the formula $(\mathbf{0} + x) = x$. Thus $\phi(\mathbf{0})$ is the formula

$$(\mathbf{0} + \mathbf{0}) = \mathbf{0} \tag{3.3}$$

and $(\phi(x) \to \phi(x'))$ is the formula

$$((\mathbf{0} + x) = x \to (\mathbf{0} + x') = x') \tag{3.4}$$

> We observed in *Unit 6*, Subsection 3.2, that the sentence $\forall x \, (\mathbf{0} + x) = x$ is not a theorem of Q, so this example shows that PA is more powerful than Q.

We show that (3.3) and (3.4) are both theorems of PA, and then, by using the appropriate induction axiom, that $\forall x \, \phi(x)$ is a theorem of PA.

1	(1)	$\forall x \, (x + \mathbf{0}) = x$	Ass
2	(2)	$\forall x \, \forall y \, (x + y') = (x + y)'$	Ass
3	(3)	$\phi(x)$	Ass
1	(4)	$\phi(\mathbf{0})$	UE, 1
2	(5)	$\forall y \, (\mathbf{0} + y') = (\mathbf{0} + y)'$	UE, 2
2	(6)	$(\mathbf{0} + x') = (\mathbf{0} + x)'$	UE, 5
2, 3	(7)	$\phi(x')$	Sub, 3, 6
2	(8)	$(\phi(x) \to \phi(x'))$	CP, 7
2	(9)	$\forall x \, (\phi(x) \to \phi(x'))$	UI, 8
10	(10)	$((\phi(\mathbf{0}) \,\&\, \forall x \, (\phi(x) \to \phi(x'))) \to \forall x \, \phi(x))$	Ass
1, 2, 10	(11)	$\forall x \, \phi(x)$	Taut, 4, 9, 10

Assumptions 1 and 2 are axioms $Q4$ and $Q5$, respectively. Assumption 10 is an induction axiom. Thus all the three assumptions on which the last line depends are axioms of PA. Hence $\vdash_{\mathrm{PA}} \forall x \, \phi(x)$, that is $\vdash_{\mathrm{PA}} \forall x \, (\mathbf{0} + x) = x$. ♦

Problem 3.1

Show that

$$\vdash_{\mathrm{PA}} \forall x\, (\mathbf{0} \cdot x) = \mathbf{0}$$

Problem 3.2

Show that

$$\vdash_{\mathrm{PA}} \forall x\, (x \cdot \mathbf{0'}) = x$$

Hint: Show that $\vdash_Q (\forall x\, (\mathbf{0} + x) = x \to \forall x\, (x \cdot \mathbf{0'}) = x)$ and then use the result of Example 3.2.

Although the theory PA is very powerful, formal proofs even of quite simple results are rather long. So we shall not try to prove many more results in PA. In fact, we shall simply show that, in the theory PA, we can prove that addition is commutative. We begin with a preliminary result.

Because of the lengths of these proofs, you may decide to take the details on trust.

Example 3.3

We show that

$$\vdash_{\mathrm{PA}} \forall x\, \forall y\, (y' + x) = (y + x)'$$

Let $\phi(x)$ be the formula $\forall y\, (y' + x) = (y + x)'$. We aim to show that both $\vdash_{\mathrm{PA}} \phi(\mathbf{0})$ and $\vdash_{\mathrm{PA}} \forall x\, (\phi(x) \to \phi(x'))$. Then a routine use of an induction axiom will enable us to deduce that $\vdash_{\mathrm{PA}} \forall x\, \phi(x)$.

Thus we show first that $\vdash_{\mathrm{PA}} \phi(\mathbf{0})$. Consider the following formal proof.

1	(1)	$\forall x\, (x + \mathbf{0}) = x$	Ass
1	(2)	$(y' + \mathbf{0}) = y'$	UE, 1
1	(3)	$(y + \mathbf{0}) = y$	UE, 1
	(4)	$(y + \mathbf{0}) = (y + \mathbf{0})$	II
1	(5)	$y = (y + \mathbf{0})$	Sub, 3, 4
1	(6)	$(y' + \mathbf{0}) = (y + \mathbf{0})'$	Sub, 2, 5
1	(7)	$\forall y\, (y' + \mathbf{0}) = (y + \mathbf{0})'$	UI, 6

The assumption on which the last line of this proof depends is axiom $Q4$. Hence the proof shows that $\vdash_{\mathrm{PA}} \phi(\mathbf{0})$. (In fact it shows that also $\vdash_Q \phi(\mathbf{0})$.)

Now consider the following formal proof.

1	(1)	$\forall x\, \forall y\, (x + y') = (x + y)'$	Ass
2	(2)	$\phi(x)$	Ass
2	(3)	$(y' + x) = (y + x)'$	UE, 2
1	(4)	$\forall y\, (z + y') = (z + y)'$	UE, 1
1	(5)	$(z + x') = (z + x)'$	UE, 4
1	(6)	$\forall z\, (z + x') = (z + x)'$	UI, 5
1	(7)	$(y' + x') = (y' + x)'$	UE, 6
1, 2	(8)	$(y' + x') = (y + x)''$	Sub, 3, 7
1	(9)	$(y + x') = (y + x)'$	UE, 6
	(10)	$(y + x') = (y + x')$	II
1	(11)	$(y + x)' = (y + x')$	Sub, 9, 10
1, 2	(12)	$(y' + x') = (y + x')'$	Sub, 8, 11
1, 2	(13)	$\phi(x')$	UI, 12
1	(14)	$(\phi(x) \to \phi(x'))$	CP, 13
1	(15)	$\forall x\, (\phi(x) \to \phi(x'))$	UI, 14

The assumption on which the last line of this proof depends is the axiom $Q5$. Hence we can deduce that $\vdash_{PA} \forall x\, (\phi(x) \to \phi(x'))$. (Again, in fact, we have $\vdash_Q \forall x\, (\phi(x) \to \phi(x')).$)

Since $\vdash_{PA} \phi(\mathbf{0})$ and $\vdash_{PA} \forall x\, (\phi(x) \to \phi(x'))$, a straightforward use of the induction axiom $((\phi(\mathbf{0})\ \&\ \forall x\, (\phi(x) \to \phi(x'))) \to \forall x\, \phi(x))$ together with the Tautology Rule enables us to deduce that $\vdash_{PA} \forall x\, \phi(x)$, that is $\vdash_{PA} \forall x\, \forall y\, (y' + x) = (y + x)'$.   ◆

Example 3.4

We show that

$$\vdash_{PA} \forall x\, \forall y\, (x + y) = (y + x)$$

Let $\phi(x)$ be the formula $\forall y\, (x + y) = (y + x)$. As in the previous example we aim to prove that both $\vdash_{PA} \phi(\mathbf{0})$ and $\vdash_{PA} \forall x\, (\phi(x) \to \phi(x'))$. We can then use an induction axiom to deduce that $\vdash_{PA} \forall x\, \phi(x)$.

By Example 3.2, $\vdash_{PA} \forall x\, (\mathbf{0} + x) = x$ and, since $Q4$ is an axiom of PA, $\vdash_{PA} \forall x\, (x + \mathbf{0}) = x$. Two applications of the UE Rule then enable us to deduce that $\vdash_{PA} (\mathbf{0} + y) = y$ and $\vdash_{PA} (y + \mathbf{0}) = y$, from which it is easy to deduce that $\vdash_{PA} (\mathbf{0} + y) = (y + \mathbf{0})$ and hence, using the UI Rule, $\vdash_{PA} \forall y\, (\mathbf{0} + y) = (y + \mathbf{0})$, that is $\vdash_{PA} \phi(\mathbf{0})$.

Now consider the following formal proof.

1	(1)	$\forall x\, \forall y\, (x + y') = (x + y)'$	Ass
2	(2)	$\forall x\, \forall y\, (y' + x) = (y + x)'$	Ass
3	(3)	$\phi(x)$	Ass
3	(4)	$(x + y) = (y + x)$	UE, 3
1	(5)	$\forall y\, (z + y') = (z + y)'$	UE, 1
1	(6)	$(z + x') = (z + x)'$	UE, 5
1	(7)	$\forall z\, (z + x') = (z + x)'$	UI, 6
1	(8)	$(y + x') = (y + x)'$	UE, 7
2	(9)	$\forall y\, (y' + z) = (y + z)'$	UE, 2
2	(10)	$(x' + z) = (x + z)'$	UE, 9
2	(11)	$\forall z\, (x' + z) = (x + z)'$	UI, 10
2	(12)	$(x' + y) = (x + y)'$	UE, 11
2, 3	(13)	$(x' + y) = (y + x)'$	Sub, 4, 12
	(14)	$(y + x') = (y + x')$	II
1	(15)	$(y + x)' = (y + x')$	Sub, 8, 14
1, 2, 3	(16)	$(x' + y) = (y + x')$	Sub, 13, 15
1, 2, 3	(17)	$\phi(x')$	UI, 16
1, 2	(18)	$(\phi(x) \to \phi(x'))$	CP, 17
1, 2	(19)	$\forall x\, (\phi(x) \to \phi(x'))$	UI, 18

Assumption 1 is axiom $Q5$, and assumption 2 is the formula which we showed in Example 3.3 is derivable in PA. Thus we can deduce that $\vdash_{PA} \forall x\, (\phi(x) \to \phi(x'))$.

Since $\vdash_{PA} \phi(\mathbf{0})$ and $\vdash_{PA} \forall x\, (\phi(x) \to \phi(x'))$, an application of an induction axiom shows that $\vdash_{PA} \forall x\, \phi(x)$, that is $\vdash_{PA} \forall x\, \forall y\, (x + y) = (y + x)$.   ◆

If you have a taste for formal proofs of this kind, you may wish to attempt the following (optional) problem.

Problem 3.3

Give an outline of a formal proof to show that

$$\vdash_{PA} \forall x\, \forall y\, \forall z\, ((y + z) + x) = (y + (z + x))$$

3.2 Gödel's Second Incompleteness Theorem

You will remember that by Hilbert's Question we mean:

Can the consistency of number theory be proved using only non-dubious principles of finitary reasoning?

To be able to answer this question, we need to make it precise in various ways. Firstly by *number theory* we shall mean, at least for the time being, the theory PA that we have just introduced.

In the previous subsection we gave a brief argument to justify the claim that PA is consistent. Roughly speaking, the argument for PA's consistency is as follows.

The axioms of PA are all true in the standard interpretation $\mathcal{N}$. Since all the rules of proof are logically valid, it follows that all the theorems of PA are true in the standard interpretation. Given a sentence Φ, it cannot be that both Φ and $\neg\Phi$ are true in the standard interpretation. So Φ and $\neg\Phi$ cannot both be theorems of PA. Hence PA is consistent.

This argument does not meet Hilbert's requirements since it takes for granted the standard interpretation of number theory and the notion of 'truth' in relation to this interpretation. But the standard interpretation is an infinite mathematical object and arguments in which we assume that we have a clear notion of truth for such infinite structures go beyond 'non-dubious principles of finitary reasoning'. Indeed Gödel's First Incompleteness Theorem shows that the properties of the standard interpretation cannot be captured by any explicitly given (that is, recursive) set of axioms in our formal language. Thus our grasp of the standard interpretation is not as comprehensive as might be suggested by the consistency proof for PA given above.

We can interpret Hilbert's requirement as asking whether we can find a recursively axiomatizable theory T which is significantly weaker than PA and in which the consistency of PA can be proved. Thus we ask whether we can find a theory T such that PA is an extension of T and a sentence which expresses the consistency of PA is a theorem of T. This immediately raises the question of how we can find a sentence of the formal language which expresses the consistency of PA. The following observation is helpful.

Theorem 3.1

Let R be a theory which extends Q. Then R is consistent if and only if the sentence $\mathbf{0} = \mathbf{1}$ is not a theorem of R.

Proof

Suppose that R is a theory which extends Q. Since $\vdash_Q \neg \mathbf{0} = \mathbf{1}$ (by Theorem 2.1(b) of *Unit 7*) and R extends Q, we also have that $\vdash_R \neg \mathbf{0} = \mathbf{1}$. Thus if $\vdash_R \mathbf{0} = \mathbf{1}$, R would be inconsistent. Conversely, if R is inconsistent, then every sentence is a theorem of R (by Problem 3.2 of *Unit 6*) and hence $\vdash_R \mathbf{0} = \mathbf{1}$. ∎

It follows that, to demonstrate the consistency of PA, all we need is to be able to express 'the sentence $\mathbf{0} = \mathbf{1}$ is not a theorem of PA' as a sentence of a recursively axiomizable theory T of which PA is an extension. We could do this with the sentence $\neg\theta(\ulcorner \mathbf{0} = \mathbf{1} \urcorner)$ if we had a formula θ, in which x is the only free variable, which expresses the fact that 'x is the Gödel number of a theorem of PA'.

Recall that if Φ is a sentence with Gödel number n, then $\ulcorner\Phi\urcorner$ is the term $\mathbf{n}$. Thus $\theta(\ulcorner\Phi\urcorner)$ is the sentence which results from substituting the term $\mathbf{n}$ for each free occurrence of x in θ.

The difficulty with this is that, by Theorem 1.3, provided PA is consistent, the set of Gödel numbers of the theorems of PA is not representable in PA. Thus we cannot expect to find a formula θ which represents the set of Gödel numbers of the theorems of PA. However, it is possible to construct a formula $\text{Prov}(x)$, in which x is the only free variable, such that, for each sentence Φ,

$\text{Prov}(\ulcorner\Phi\urcorner)$ is true in the standard interpretation if and only if $\vdash_{\text{PA}} \Phi$.

The construction of the formula Prov involves a good deal of technical work, which we omit. We hope that you will take the existence of Prov on trust.

Given the formula Prov, we can restate Hilbert's Question as follows:

Is there a recursively axiomatizable theory T such that PA is an extension of T and $\vdash_T \neg\text{Prov}(\ulcorner\mathbf{0 = 1}\urcorner)$?

We shall show that the answer to this question is 'no' by proving that, if PA is consistent, then $\neg\text{Prov}(\ulcorner\mathbf{0 = 1}\urcorner)$ is not even a theorem of PA. It then immediately follows that it is not a theorem of any theory T of which PA is an extension.

Rather than prove this for the particular formula Prov, it is convenient to prove this result in a more general setting (in Theorem 3.3) which deals with any formula that might be regarded as expressing that formulas are provable in PA. We call such formulas *provability predicates*.

Definition 3.2 Provability Predicate

Let T be a theory which is an extension of Q and let θ be a formula whose only free variable is x. Then θ is called a *provability predicate* for T if the following three conditions are satisfied for all sentences Φ and Ψ.

(a) If $\vdash_T \Phi$ then $\vdash_T \theta(\ulcorner\Phi\urcorner)$.

(b) $\vdash_T (\theta(\ulcorner\Phi \to \Psi\urcorner) \to (\theta(\ulcorner\Phi\urcorner) \to \theta(\ulcorner\Psi\urcorner)))$.

(c) $\vdash_T (\theta(\ulcorner\Phi\urcorner) \to \theta(\ulcorner\theta(\ulcorner\Phi\urcorner)\urcorner))$.

If we think of $\theta(\ulcorner\Phi\urcorner)$ as expressing the fact that 'Φ is a theorem of T', we can see that the conditions in Definition 3.2 express the following statements.

(a) If Φ is a theorem of T, then we can prove in T that Φ is a theorem of T.

(b) We can prove in T that, if $(\Phi \to \Psi)$ and Φ are theorems of T, then Ψ is a theorem of T.

(c) We can prove in T that, if Φ is a theorem of T, then we can prove in T that Φ is a theorem of T.

The following theorem, interesting in its own right, which will give us our final stepping stone to answering Hilbert's Question, is proved using another very ingenious application of Gödel's Diagonal Lemma. The individual steps in the proof are not difficult to follow. They use the properties (a), (b) and (c) of a provability predicate and the Tautology Rule. The tautologies used are the following, expressed in the notation of Subsection 1.3 of *Unit 6*.

(i) $\dfrac{\phi}{(\psi \to \phi)}$ (ii) $\dfrac{(\phi \to \psi), \phi}{\psi}$ (iii) $\dfrac{(\phi \to \psi), (\psi \to \chi)}{(\phi \to \chi)}$

(iv) $\dfrac{(\phi \to (\psi \to \chi)), (\phi \to \psi)}{(\phi \to \chi)}$ (v) $\dfrac{(\phi \leftrightarrow \psi), \psi}{\phi}$

Tautologies (ii) and (iii) are the standard tautological consequences 8 and 10 given in Subsection 1.3 of *Unit 6*.

We realize that the proof as a whole is rather difficult to grasp, so you should not worry if you find it hard to see what is really going on. We shall not expect you to remember either the theorem or its proof, but we have included it so that you can see what is involved in answering Hilbert's Question.

> ### Theorem 3.2 Löb's Theorem
>
> Let T be a theory which extends Q and let θ be a provability predicate for T. Then, for each sentence Φ, $\vdash_T \Phi$ if and only if $\vdash_T (\theta(\ulcorner\Phi\urcorner) \to \Phi)$.

This theorem was proved by M.H. Löb in 1955.

Proof

Suppose that T is a theory which extends Q and that θ is a provability predicate for T. Let Φ be a sentence.

If $\vdash_T \Phi$ then by the Tautology Rule (using tautology (i)) $\vdash_T (\theta(\ulcorner\Phi\urcorner) \to \Phi)$.

Conversely, suppose that

$$\vdash_T (\theta(\ulcorner\Phi\urcorner) \to \Phi) \tag{3.5}$$

Let $\delta(x)$ be the formula $(\theta(x) \to \Phi)$ in which x is the only variable which may occur freely. Since T is an extension of Q, the function diag is representable in T, and so we can apply Gödel's Diagonal Lemma (Theorem 3.5 of *Unit 7*) to deduce that there is a sentence G_δ such that $\vdash_T (G_\delta \leftrightarrow \delta(\ulcorner G_\delta\urcorner))$, that is

We met the function diag and saw that it is representable in any theory that extends Q in Subsection 3.2 of *Unit 7*.

$$\vdash_T (G_\delta \leftrightarrow (\theta(\ulcorner G_\delta\urcorner) \to \Phi)) \tag{3.6}$$

and hence

$$\vdash_T (G_\delta \to (\theta(\ulcorner G_\delta\urcorner) \to \Phi)) \tag{3.7}$$

Using property (a) of a provability predicate, (3.7) gives

$$\vdash_T \theta(\ulcorner(G_\delta \to (\theta(\ulcorner G_\delta\urcorner) \to \Phi))\urcorner) \tag{3.8}$$

Now, by property (b) of a provability predicate, we have

$$\vdash_T (\theta(\ulcorner(G_\delta \to (\theta(\ulcorner G_\delta\urcorner) \to \Phi))\urcorner) \to (\theta(\ulcorner G_\delta\urcorner) \to \theta(\ulcorner(\theta(\ulcorner G_\delta\urcorner) \to \Phi)\urcorner))) \tag{3.9}$$

The Tautology Rule applied to (3.8) and (3.9) (tautology (ii)) gives

$$\vdash_T (\theta(\ulcorner G_\delta\urcorner) \to \theta(\ulcorner(\theta(\ulcorner G_\delta\urcorner) \to \Phi)\urcorner)) \tag{3.10}$$

Using property (b) again, we have

$$\vdash_T (\theta(\ulcorner(\theta(\ulcorner G_\delta\urcorner) \to \Phi)\urcorner) \to (\theta(\ulcorner\theta(\ulcorner G_\delta\urcorner)\urcorner) \to \theta(\ulcorner\Phi\urcorner))) \tag{3.11}$$

The Tautology Rule applied to (3.10) and (3.11) (tautology (iii)) gives

$$\vdash_T (\theta(\ulcorner G_\delta\urcorner) \to (\theta(\ulcorner\theta(\ulcorner G_\delta\urcorner)\urcorner) \to \theta(\ulcorner\Phi\urcorner))) \tag{3.12}$$

By property (c) of a provability predicate, we have

$$\vdash_T (\theta(\ulcorner G_\delta\urcorner) \to \theta(\ulcorner\theta(\ulcorner G_\delta\urcorner)\urcorner)) \tag{3.13}$$

The Tautology Rule applied to (3.12) and (3.13) (tautology (iv)) gives

$$\vdash_T (\theta(\ulcorner G_\delta\urcorner) \to \theta(\ulcorner\Phi\urcorner)) \tag{3.14}$$

Now the Tautology Rule applied to (3.5) and (3.14) (tautology (iii)) gives

$$\vdash_T (\theta(\ulcorner G_\delta\urcorner) \to \Phi) \tag{3.15}$$

We can now apply the Tautology Rule to (3.6) and (3.15) (tautology (v)) to deduce that

$$\vdash_T G_\delta \tag{3.16}$$

Hence, by property (a) of a provability predicate, (3.16) gives

$$\vdash_T \theta(\ulcorner G_\delta\urcorner) \tag{3.17}$$

Finally from (3.15) and (3.17) and another application of the Tautology Rule (tautology (ii)), we can deduce that

$$\vdash_T \Phi \tag{3.18}$$

We have thus shown that (3.5) implies (3.18), completing the proof. ∎

This theorem leads immediately to the result we need.

Theorem 3.3

If T is a consistent extension of Q and θ is a provability predicate for T then $\neg\theta(\ulcorner \mathbf{0} = \mathbf{1} \urcorner)$ is not a theorem of T.

Proof

Suppose that T is a consistent extension of Q and that θ is a provability predicate for T.

Now assume that

$$\vdash_T \neg\theta(\ulcorner \mathbf{0} = \mathbf{1} \urcorner)$$

By an application of the Tautology Rule, using the tautology $(\neg\phi \rightarrow (\phi \rightarrow \psi))$, it follows that

$$\vdash_T (\theta(\ulcorner \mathbf{0} = \mathbf{1} \urcorner) \rightarrow \mathbf{0} = \mathbf{1})$$

Hence, by Theorem 3.2,

$$\vdash_T \mathbf{0} = \mathbf{1}$$

Now, as T is an extension of Q, it follows, by Theorem 3.1, that T is not consistent. From this contradiction we can deduce that $\neg\theta(\ulcorner \mathbf{0} = \mathbf{1} \urcorner)$ cannot be a theorem of T. ∎

We wish to use Theorem 3.3 to show that, if PA is consistent, then $\neg\mathrm{Prov}(\ulcorner \mathbf{0} = \mathbf{1} \urcorner)$ is not a theorem of PA. To be able to draw this conclusion, we need to know that Prov is a provability predicate for PA. The detailed verification of this is complicated, so we hope you will take it on trust that

Prov is a provability predicate for PA.

It is worth remarking that here we really do need a powerful theory with lots of axioms such as PA. For Prov to be a provability predicate, we need to know that conditions (a), (b) and (c) of Definition 3.2 hold, and this requires that the sentences given in these conditions are theorems of our theory. For this to be true, a powerful theory is needed. The analogous result for a weak theory such as Q would not hold.

Theorem 3.4 Gödel's Second Incompleteness Theorem

If Peano Arithmetic, PA, is consistent, then $\neg\mathrm{Prov}(\ulcorner \mathbf{0} = \mathbf{1} \urcorner)$ is not a theorem of PA.

Proof

Since PA is an extension of Q, and Prov is a provability predicate for PA, this is an immediate consequence of Theorem 3.3. ∎

We have included the hypothesis that PA is consistent in our statement of Theorem 3.4 since its main interest stems from Hilbert's Question, and this question has little significance if the consistency of PA is taken for granted. However, if we do accept the existence of the standard interpretation $\mathcal{N}$ and hence the consistency of PA, then $\neg\mathrm{Prov}(\ulcorner \mathbf{0} = \mathbf{1} \urcorner)$ is a true sentence (in the standard interpretation). Thus $\neg\mathrm{Prov}(\ulcorner \mathbf{0} = \mathbf{1} \urcorner)$ is an example of a true sentence that is not a theorem of PA. Since the sentence $\mathrm{Prov}(\ulcorner \mathbf{0} = \mathbf{1} \urcorner)$ is false, then it is also not a theorem of PA. So we obtain an example of a sentence Φ such that neither $\vdash_{PA} \Phi$ nor $\vdash_{PA} \neg\Phi$ and thus a direct example to show that the theory PA is not complete. This is essentially the proof that Gödel gave of his First Incompleteness Theorem.

Now we are in a position to answer Hilbert's Question, at least in the way we have reformulated it. Taking together Theorems 3.1 and 3.4 we can conclude that

if PA is consistent, then we cannot prove in PA that it is consistent.

So our answer to Hilbert's Question is 'no'.

Despite this answer, since almost all mathematicians believe that they have a clear and coherent concept of the natural numbers and hence of what we have called the standard interpretation, the consistency of the theory PA is not seriously in doubt. However, note that Theorem 3.3 applies to all theories which extend PA, and this includes *theories of infinite sets* which are very much more powerful than PA. For these powerful set theories there is no standard interpretation, or indeed any other interpretation, of which it is plausible to claim that we have a very clear conception. Thus the possibility that such theories are inconsistent needs to be taken seriously. The absence of finitary consistency proofs for such theories therefore has great philosophical significance. It shows that Hilbert's idea for securing the consistency of the mathematics of the infinite cannot be accomplished.

BIOGRAPHICAL SKETCHES

In this section we give some brief biographical sketches of some of the mathematicians and logicians we have mentioned earlier in the course.

Cantor

Georg Cantor (1845–1918) was born to German parents in St Petersburg, Russia. His father was a successful Protestant merchant, his mother a Catholic; religion was to play an increasing role in Cantor's life as he grew older. He studied mathematics at Berlin University, where he was attracted to the rigorous approach to analysis advocated by Karl Weierstrass.

His first significant research was on the convergence of trigonometric series, where he was led to the delicate study of point sets on the real line, which later influenced his ideas about ordinal numbers. In 1871 he began a long correspondence with Richard Dedekind, in which he showed that the rational numbers formed a countable set but the real numbers formed a strictly larger, uncountable set. Since Cantor had also shown that the set of algebraic numbers is countable, this immediately implied the existence of uncountably many transcendental numbers.

Cantor went on to explore the new world of uncountable sets. He showed, for example, that there was a one-to-one correspondence between the points of an interval and a square, which seemed to imperil the concept of dimension, and was led to conjecture that any infinite subset of the real numbers can be put in a one-to-one correspondence with either the set of natural numbers or the set of real numbers. This is the famous continuum hypothesis, unproved to this day, despite modern set theory being considerably more precise than it was in Cantor's time.

Cantor may also have been the first to discover the paradoxes of the largest cardinal and the largest ordinal. The first follows from the facts that every set has a cardinality, and the 'set' of all cardinals has a cardinality larger than any cardinal. The paradox of the 'set' of all ordinals is similar. The resolution of the paradoxes, when they began to cause concern in the 1900s, has generally been taken to lie in realizing that not all collections can form sets; but Cantor seems to have regarded them with equanimity because they appealed to his ideas about the truly infinite and the nature of God.

Cantor spent all his working life as a professor at the small University of Halle, hoping for a call that never came from the University of Berlin. He seems to have been devoted to his family, but suffered a number of periods of acute mental illness. Cantor's work did not get the recognition it deserved from some of his contemporaries, who did not accept the validity of his theory of infinite sets. However, the younger generation of mathematicians, led by David Hilbert, regarded set theory as a mathematical paradise and the proper place for the foundation of all mathematics.

Church

Alonzo Church (1903–1995) was born in Washington DC in the USA and spent much of his professional life at Princeton University, where he obtained both his bachelor and PhD degrees and then served as a professor from 1929 to 1967. He produced work of major importance in logic, recursion theory and computer science, publishing his first paper in 1924 and his last in 1995, the year he died. His achievements include his work on recursive functions leading to his proposal of what became known as Church's Thesis, his proof that there is no algorithm for deciding which sentences of quantifier logic are logically valid, and his creation of the lambda calculus. He was the doctoral supervisor of many of the most distinguished logicians of the twentieth century, including Stephen Kleene and Alan Turing. He was known for his extreme precision when lecturing. His *Introduction to Mathematical Logic*, published in 1956, which also shows great care with the details, has proved to be a very influential textbook. He further fostered the study of logic as co-founder of the Association of Symbolic Logic in 1936 and by playing a leading role in running for many years the *Journal of Symbolic Logic*, the premier learned journal devoted to mathematical logic.

Alonzo Church (Photo © the estate of Alonzo Church)

Frege

Gottlob Frege (1848–1925) became a student at Jena University in 1869. He moved to Gottingen University in 1871 and was awarded a PhD in 1873. He then returned to Jena University where he taught mathematics for forty-four years and was known as clear, conscientious and demanding teacher. His research concentrated more and more on the foundations of mathematics. Frege came to believe that the basic concepts of mathematics, with the possible exception of geometry, could be defined in purely logical terms, and theorems could then be derived using just logical reasoning. To justify this claim, in his *Begriffsschrift* (Conceptual Notation) of 1879 he described a formal system in which deductions were carried out. This was the first published system of quantifier logic and Frege merits great credit for his invention. Unfortunately he chose an awkward notation and this restricted the influence of his book.

After describing his philosophical ideas informally in *Die Grundlagen der Arithmetik* (The Foundations of Mathematics) in 1884, he began work on *Die Grundgesetze der Arithmetik* (The Basic Laws of Arithmetic), which was intended to be his definitive account of how theorems of number theory and mathematical analysis may be derived within his formal system. The first volume was published in 1893. However, in June 1902, just as the second volume was nearing completion, he received a letter from Bertrand Russell drawing his attention to a contradiction, now called Russell's Paradox, which Russell had derived within Frege's system. To his great credit, Frege, who had indulged in polemics much of his working life, gracefully acknowledged the force of Russell's criticism. He was never able to repair the damage and towards the end of his life he seems to have abandoned his belief that number could be derived purely from logic.

Gottlob Frege (Photo courtesy of the Friedrich-Schiller-University, Jena)

Frege's work won little attention in his lifetime, but he is now regarded as a founder of mathematical logic and as a great philosopher of mathematics and logic.

Gödel

Kurt Gödel (1906–1978), who was a native of Brno (in modern-day Czech Republic), who studied mathematics at the University of Vienna and took his PhD there in 1930. One of his main teachers of mathematics was Hans Hahn whose interests included point-set topology, the foundations of mathematics and the philosophy of science. Hahn introduced Gödel to a group of positivistic philosophers now known as the Vienna Circle. Gödel attended their meetings but never accepted their philosophical ideas. His first achievement was his theorem, which we have called the Adequacy Theorem, showing that every logically valid formula in first-order logic is derivable in a formal proof system. He first presented his First Incompleteness Theorem at a conference in Königsberg when he was 24.

With the rise of the Nazis it was prudent for him to emigrate, and he eventually reached America, via Russia, in 1940, when he took up a permanent position at the recently founded Institute for Advanced Study in Princeton. There he remained for the rest of his life. He worked at first on set theory and in 1938 published a proof that if the standard axiom system for set theory is consistent then Cantor's continuum hypothesis cannot be disproved in the system. He became a good friend of Albert Einstein, who was also a professor at the Institute, and even discovered a novel space-time that seems to allow the possibility of time travel. In his final years he became a complete recluse, and starved to death in 1978 because of his fear of consuming harmful germs.

Kurt Gödel (Photo courtesy of the Archives of the Institute for Advanced Study)

Hilbert

David Hilbert (1862–1943) was born and studied in Königsberg in East Prussia and became a professor there in 1892. His transfer to Göttingen in 1895 marked the rise of that university to become the leading centre for mathematics in the world. Hilbert's remarkable mathematical ability showed itself in the late 1880s when he solved a famously difficult problem in algebra, making great use of a non-constructive existence argument (which proved to be a favourite method of his). He worked on numerous branches of mathematics: number theory, functional analysis, and mathematical physics including general relativity. He was a passionate advocate of Cantorian set theory, and of the axiomatic approach to pure, and applied, mathematics, which he first developed in the context of elementary geometry. In the second half of his career, he hoped to create a combined version of logic and set theory from which could be derived all of mathematics in a rigorous way from simple logical foundations. Although he failed in this, many of the fragments of his achievement remained for later mathematical logicians to refine.

Königsberg is today called Kaliningrad and is in Russia.

Kleene

Stephen Kleene (1909–1994) — the name is pronounced clay-nee — was born in Hartford, Connecticut, USA, and took his PhD at Princeton University in 1934. He was a founder of modern recursion theory, and in 1983 he was awarded the American Mathematical Society's Steele Prize for his seminal papers of 1955 on recursion theory and descriptive set theory. He taught for most of his life at the University of Wisconsin–Madison, where he built up a large school of logicians in the Mathematics Department. Among his influential papers and books are his *Introduction to Metamathematics* (1952) and *Mathematical Logic* (1962). In 1990 Kleene was awarded the President's National Medal of Science, America's highest scientific honour, for his leadership in the theory of recursion and effective computability and for developing it into a deep and broad field of mathematical research. He was also a keen mountaineer and an energetic conservationist.

Leibniz

Gottfried Wilhelm Leibniz (1646–1716) was, with Isaac Newton, the commanding mathematician of his time, though most of his working life was spent working for the Duke of Brunswick on a genealogy project that prevented his remarkable talents from exerting their full influence. His own discovery of the fundamental ideas of the calculus came in the late 1670s, some years after Newton's discovery, which was at that time unpublished. Leibniz's interests in philosophy were profound, and he also held out hopes for a universal language and for a language in which all ideas could be expressed and all disputes resolved by calculation. He could not match Newton when it came to applying mathematics to the study of nature, but he surpassed him in extending the techniques of the calculus and in passing it to the next generation through his correspondence with the Bernoulli brothers Johann and Jakob. In the late nineteenth century many of his hitherto unpublished ideas on logic and metaphysics were made available for the first time in scholarly editions, and became a rallying point for Frege, Russell and others.

Peano

Giuseppe Peano (1858–1932) was an Italian mathematician who taught almost all his life at the University of Turin. He opposed the sweeping use of geometrical arguments and instead preferred careful analytic arguments often tending to seemingly paradoxical conclusions, such as his discovery of a continuous map of the unit interval onto the unit square, the first example of a space-filling curve. He proceeded to introduce much of the now-standard notation in logic and set theory, which is what commended him to Bertrand Russell, and to advocate a form of simplified Latin for use in all scientific work (and indeed, more generally, as an alternative to Esperanto). For a time he inspired a group of Italian mathematicians and logicians, but after the First World War he was an increasingly isolated figure.

Apart from his other contributions to mathematics, Peano is renowned for his skill in devising good notation. It is Peano to whom we are indebted for, for instance, the symbol $\in$ for set membership. In his autobiography, Bertrand Russell wrote that the International Congress of Philosophy in Paris in 1900 'was a turning point in my intellectual life, because I there met Peano .... It became clear to me that his notation afforded an instrument of logical analysis such as I had been seeking for years, and that by studying him I was acquiring a new and powerful technique for the work I had long wanted to do.'

Post

Emil Post (1897–1954) was born into a Jewish family in Poland in 1897 and emigrated with his parents to New York in 1904. He studied mathematics at City College and took his PhD from Columbia University, New York, in 1920. He lost an arm in an accident as a child, and suffered all his adult life from crippling manic depression, which prevented him from obtaining a permanent position until 1935 when he became a professor at City College, where he remained until he died in 1954. Post first studied logic at a seminar at Columbia University as a graduate student. In his thesis he isolated the propositional logic part of Russell and Whitehead's *Principia Mathematica* and used the method of truth tables to show it was consistent and complete. He regarded systems of logic as systems of finitary symbol manipulation, and in later work came close to Gödel's Incompleteness Theorems. In 1943 he lectured to the American Mathematical Society on the field of recursive unsolvability — probably his most influential achievement — and his theorems in this area have inspired other proofs of unsolvability in the theory of formal languages.

Russell

Bertrand Russell (1872–1970) was born into an aristocratic family. He studied mathematics and philosophy as a student at Trinity College, Cambridge. He won a fellowship in 1895 for an essay subsequently published as *An Essay on the Foundations of Geometry*. His interest in the philosophy of mathematics was enhanced by a meeting with Giuseppe Peano in 1900. Independently of Gottlob Frege, Russell came to the view that mathematics could be reduced to logic. This thesis is put forward in his *Principles of Mathematics* which he began writing in 1900 but which was not published until 1903. During this period he got to know of Frege's work, and he also discovered the paradox now known as Russell's Paradox.

Russell planned to follow *Principles of Mathematics* with a second volume in which his claim that all the theorems of pure mathematics are deducible using 'a small number of fundamental logic principles' would be established 'by strict symbolic reasoning'. To achieve this aim Russell collaborated with the Cambridge mathematician Alfred North Whitehead, and their *Principia Mathematica* was published in three volumes between 1910 and 1913. In this work Russell's Paradox was avoided by using Russell's Theory of Types. This makes the logical system complicated and undermines the claim that mathematics has been deduced from 'small number of fundamental logical principles'.

Russell courageously opposed the First World War. His political activities led to the loss in 1916 of his lectureship at Trinity College, and to his being jailed in 1918 for publishing a seditious pamphlet. After publishing *Principia Mathematica* he only rarely returned to serious work in mathematical logic. Towards the end of his life he became active in the campaign against nuclear weapons and in 1961 he was imprisoned again for short period.

Bertrand Russell (Photo © Mary Evans Picture Library)

Turing

Alan Mathison Turing (1912–1954) studied mathematics at King's College, Cambridge, where he developed an interest in logic. In 1936 he gave the first proof of an algorithmically unsolvable problem when he showed that the Halting Problem is algorithmically undecidable (Theorem 3.1 of *Unit 3*). In the course of this work he devised the notion of a Turing machine, a particularly clear way of describing an automated reasoning process. During the Second World War he was very influential in the code-breaking efforts centred at Bletchley Park, and this led to work on electronic computers that he later continued at the University of Manchester. In March 1952 he was charged with committing homosexual acts which were then illegal. He pleaded guilty and was put on probation subject to the condition that he undertook a course of hormone injections that were supposed to reduce his homosexual urges. He committed suicide on 7 June 1954 by eating an apple laced with cyanide.

SUGGESTIONS FOR FURTHER READING

Because of the great significance of Gödel's Incompleteness Theorems, the literature on the subject is vast. We give only a selection of books. They are a mixture of works of historical or biographical interest and some textbooks which present the proofs of Gödel's theorems and related results.

With the exception of Gödel's original paper, we restrict ourselves to works in English. There are lots of different, but equivalent, ways of presenting systems of formal logic. So if you look at any of the books listed here, you must be prepared for differences in notation and presentation.

The details of the books mentioned are in the bibliography given at the end. We give references to this bibliography in the form [n], where n is the number of the book in the bibliography.

Most of these books should be available in a good academic library. Public libraries should be able to obtain them through the inter-library loan system.

Recursive functions, Turing machines and URMs

Alan Turing's original paper is [29]. This is reprinted in [7] and [13]. The original paper in which URMs were described is [25]. Our treatment of URMs follows very closely that of Nigel Cutland [4]. Note, however, that we have used $C(m, n)$ for Copy instructions whereas Cutland uses $T(m, n)$. Also we have used a different system for assigning code numbers to URM programs and URM computations so as to bring it into line with our method of assigning Gödel numbers to formulas. For a detailed treatment of computable functions in terms of Turing machines see [6]. Recursive functions are also covered in [10], [16], [18] and [19].

Formal systems of logic

Church's book [3] is a classic text which covers propositional and quantifier logic in great detail, but which covers neither recursive functions nor Gödel's theorems (a promised second volume covering them never appeared). Kleene's book [18] is another classic text which does include recursive functions and Gödel's theorems.

Both these influential textbooks are hard going for beginners. There is now available a large number of more accessible introductions to mathematical logic. We have chosen to recommend [10], [16] and [19]. Each of these books has stood the test of time. They all give a more detailed introduction to propositional and quantifier logic than we have had space for. In particular they all give proofs of the Correctness Theorem and the Adequacy Theorem for quantifier logic, which we omitted. They also cover recursive functions and Gödel's Incompleteness Theorems, and give proofs of the representability of total recursive functions in sufficiently powerful systems of formal arithmetic, which we did not give in full.

The Correctness Theorem is called the Soundness Theorem in [10] and [16] and is unnamed in [19]. The Adequacy Theorem is called the Completeness Theorem in [10] and [19].

Gödel's Incompleteness Theorems

Gödel's original paper is [14]. English translations are given in [15] (which also includes the original German text), [7], [24] and [30]. Each of these volumes is also of separate interest.

The first volume of Gödel's *Collected Works* [15] includes a biographical sketch of Gödel by John Dawson and a useful introductory note about Gödel's paper by Kleene. [7] is a collection of reprints of some important papers which also includes Turing's paper [29] and the paper in which Church first put forward what we have called Church's Thesis. The volume [24] edited by S.G. Shanker includes a selection of philosophical and historical papers commenting on the reception of Gödel's theorems and their philosophical significance. The volume [30] edited by Jean van Heijenoort is a very scholarly volume consisting of key papers in mathematical logic with useful editorial introductions.

An advanced, but none the less very readable, book illustrating further some of the strong connections between the number theory and mathematical logic encountered in this course is [26]. Another advanced but still very readable book on computability and logic, giving further interesting historical background, is [11].

Two amusing and challenging books of logic puzzles by the logician (and magician) Raymond Smullyan, which lead to discussions of Gödel's theorems and what we have called Leibniz's Question, are [27] and [28].

Biographies

Martin Davis's book [8] is an entertaining short history describing some of the pathway connecting the ideas of Leibniz with the work of Gödel and Turing. The book by E.J. Aiton [2] is an accessible biography of Leibniz. Probably the most accessible source of biographical information about Cantor is [1]. J.W. Dauben's book [5] is a more scholarly account of Cantor's work with some biographical information. Many of Frege's papers were destroyed during the Second World War and as a consequence many details of his life are not known. Our biographical comments are based on Terrell Ward Bynum's biography in [12]. This volume also includes a translation of Frege's *Begriffsschrift*, as also does [30]. Constance Reid's book [22] is a very readable account of the life and work of Hilbert. The lives of Gödel and Turing are covered in [9] and [17]. Russell wrote an interesting but not always reliable or balanced autobiography [23]. He has attracted the interest of several biographers: the two-volume biography by Ray Monk [20], [21] is perhaps the best, and not always favourable to Russell.

Bibliography

The dates given are normally those of the first edition of each book, many of which have been subsequently reprinted.

1. Amir D. Aczel, *The Mystery of the Aleph*, Washington Square Press, 2000.

2. E.J. Aiton, *Leibniz: a Biography*, Adam Hilger, 1985.

3. Alonzo Church, *Introduction to Mathematical Logic, volume 1*, Princeton University Press, 1956.

4. Nigel Cutland, *Computability*, Cambridge University Press, 1980.

5. J.W. Dauben, *Georg Cantor*, Princeton University Press, 1990.

6. Martin Davis, *Computability and Unsolvability*, McGraw-Hill, 1958.

7. Martin Davis (editor), *The Undecidable*, Raven Press, 1965.

8. Martin Davis, *The Universal Computer: the Road from Leibniz to Turing*, W.W. Norton, 2000.

9. John Dawson, *Logical Dilemmas: the Life and Work of Kurt Gödel*, A.K. Peters, 1997.

10. Herbert B. Enderton, *A Mathematical Introduction to Logic*, second edition, Academic Press, 2001.

11. R.L. Epstein and W.A. Carnielli, *Computability: Computable Functions, Logic and the Foundations of Mathematics*, Wadsworth, 1989.

12. Gottlob Frege, *Conceptual Notation and Related Articles*, translated and edited with a Biography and Introduction by Terrell Ward Bynum, Clarendon Press, 1972.

13. R.O. Gandy and C.E.M. Yates (editors), *Collected Works of A.M. Turing: Mathematical Logic*, Elsevier, 2001.

14. Kurt Gödel, 'Über formal unentscheidbare Sätze per Principia Mathematica und verwandter Systeme I', *Monatshefte für Mathematik und Physik*, volume 38, 1931, pages 173–198.

15. Kurt Gödel, *Collected Works, volume 1*, edited by Solomon Feferman *et al*, Oxford University Press, 1986.

16. A.G. Hamilton, *Logic for Mathematicians*, Cambridge University Press, 1978.

17. Andrew Hodges, *Alan Turing: the Enigma*, Burnett Books, 1983.

18. Stephen Cole Kleene, *Introduction to Metamathematics*, North-Holland, 1952

19. Elliott Mendelson, *Introduction to Mathematical Logic*, Van Nostrand, 1964.

20. Ray Monk, *Bertrand Russell: The Spirit of Solitude*, Jonathan Cape, 1996.

21. Ray Monk, *Bertrand Russell: The Ghost of Madness*, Jonathan Cape, 2000.

22. Constance Reid, *Hilbert*, George Allen & Unwin/Springer Verlag, 1970.

23. Bertrand Russell, *The Autobiography of Bertrand Russell*, three volumes, George Allen & Unwin, 1967, 1968 and 1969.

24. S.G. Shanker (editor), *Gödel's Theorem in Focus*, Croom Helm, 1988.

25. J.C. Shepherdson and H.E. Sturgis, 'Computability of recursive functions', *Journal of the Association of Computing Machinery*, volume 10, 1963, pages 217–255.

26. Craig Smoryński, *Logical Number Theory, volume 1*, Springer Verlag, 1991.

27. Raymond Smullyan, *What is the Name of this Book?*, Pelican, 1981.

28. Raymond Smullyan, *The Lady or the Tiger*, Pelican, 1983.

29. A.M. Turing, 'On computable numbers, with an application to the Entscheidungsproblem', *Proceedings of the London Mathematical Society*, series 2, volume 42, 1936–37, pages 230–265.

30. Jean van Heijenoort (editor), *From Frege to Gödel: a Source Book in Mathematical Logic*, Harvard University Press, 1967.

SUMMARY

In this unit we obtained answers to the questions which we posed at the beginning of *Unit 1*, Leibniz's Question and Hilbert's Question.

First we exploited Gödel's Diagonal Lemma to prove that, for any consistent theory T which extends Q, the set $GN(T)$ of Gödel numbers of theorems of T is not representable in T. We then exploited this result to show that the set of Gödel numbers of the sentences true in the standard interpretation $\mathcal{N}$ is not recursive. Given the evidence for Church's Thesis, which identifies the notions of algorithmically computable function and recursive function, this means that there is no algorithm for deciding which statements of number theory are true, thus providing a negative answer to Leibniz's Question.

We then asked whether there is a set of axioms for the theory Complete Arithmetic (CA) for which there is an algorithm to decide whether any given formula is one of the axioms. The existence of such a set of axioms would have led to a way of deciding whether a given sequence of formulas was a correct formal proof within the theory CA. This led to a discussion of the ideas of a *recursively axiomatizable* theory and a *recursively enumerable* set of natural numbers. Using these ideas we proved Gödel's First Incompleteness Theorem which states that there is no complete and consistent recursively axiomatizable extension of Q. Thus, in particular, CA is not recursively axiomatizable.

Finally we introduced the theory *Peano Arithmetic* (PA) which is recursively axiomatizable and from which a significant portion of number theory can be derived. We discussed the idea of a *provability predicate* of a theory. Using, in particular, the provability predicate Prov for PA we proved Gödel's Second Incompleteness Theorem which shows that the sentence $\neg\text{Prov}(\ulcorner \mathbf{0} = \mathbf{1} \urcorner)$ is not a theorem of PA. In the standard interpretation this sentence essentially asserts that the theory PA is consistent. The impossibility of deriving this sentence in the theory PA gives a negative answer to Hilbert's Question.

OBJECTIVES

We list topics on which we may set examination questions to test your understanding of this unit.

After working through the unit you should be able to explain the meaning of each of the following results (and thus of the technical terms used in stating them) and how they relate to Leibniz's and Hilbert's Questions.

(a) *Theorem 1.3*
Let T be a consistent theory which extends Q. Then the set $GN(T)$ of Gödel numbers of the theorems of T is not representable in T.

(b) *Theorem 1.4 Negative Answer to Leibniz's Question*
The set of Gödel numbers of the sentences which are true in the standard interpretation $\mathcal{N}$ is not recursive, or equivalently there is no algorithm for deciding which statements of number theory are true.

(c) *Theorem 1.5*
Let T be a consistent theory which extends Q. Then the set of the Gödel numbers of the sentences which are theorems of T is not recursive.

(d) *Theorem 1.6*
There is no algorithm for deciding which sentences are theorems of Q.

(e) *Theorem 1.9 Church's Theorem*
There is no algorithm for deciding which sentences of quantifier logic are logically valid.

(f) *Theorem 2.1*
The set $GN(T)$ of Gödel numbers of the theorems of a recursively axiomatizable theory T form a recursively enumerable set.

(g) *Theorem 2.2*
A complete recursively axiomatizable theory has a recursive set of theorems.

(h) *Theorem 2.3 Gödel's First Incompleteness Theorem*
There is no complete and consistent recursively axiomatizable extension of Q.

(i) *Theorem 2.4*
Complete Arithmetic is not recursively axiomatizable.

(j) *Theorem 3.4 Gödel's Second Incompleteness Theorem*
If Peano Arithmetic, PA, is consistent, then $\neg\mathrm{Prov}(\ulcorner \mathbf{0} = \mathbf{1} \urcorner)$ is not a theorem of PA, thus providing a negative answer to Hilbert's Question.

We shall not test your understanding of the *proofs* of these theorems.

SOLUTIONS TO THE PROBLEMS

Solution 1.1

Suppose that δ_A is a formula with x as its only free variable, such that (a) if $n \in A$ then $\vdash_T \delta_A(\mathbf{n})$ and (b) if $n \notin A$ then $\vdash_T \neg\delta_A(\mathbf{n})$.

Let ϕ_A be the formula

$$((\delta_A(x) \mathbin{\&} y = \mathbf{1}) \vee (\neg\delta_A(x) \mathbin{\&} y = \mathbf{0}))$$

We show that ϕ_A represents the characteristic function χ_A of A in the theory T.

Take $n, k \in \mathbb{N}$ and suppose that $\chi_A(n) = k$.

We consider first the case where $n \in A$, in which case $k = 1$.

Since $n \in A$, $\vdash_T \delta_A(\mathbf{n})$. By the II Rule $\vdash_T \mathbf{1} = \mathbf{1}$. Hence, using the Tautology Rule, it follows that

$$\vdash_T ((\delta_A(\mathbf{n}) \mathbin{\&} \mathbf{1} = \mathbf{1}) \vee (\neg\delta_A(\mathbf{n}) \mathbin{\&} \mathbf{1} = \mathbf{0}))$$

that is $\vdash_T \phi_A(\mathbf{n}, \mathbf{1})$. Hence, as $k = 1$, we have

$$\vdash_T \phi_A(\mathbf{n}, \mathbf{k})$$

and so the first condition for representability is satisfied by ϕ_A when $n \in A$.

Since $\vdash_T \delta_A(\mathbf{n})$, using the Tautology Rule we have that

$$\vdash_T (((\delta_A(\mathbf{n}) \mathbin{\&} y = \mathbf{1}) \vee (\neg\delta_A(\mathbf{n}) \mathbin{\&} y = \mathbf{0})) \to y = \mathbf{1})$$

> Here we use the fact that the formula
> $(\gamma \to (((\gamma \mathbin{\&} \psi) \vee (\neg\gamma \mathbin{\&} \chi)) \to \psi))$
> is a tautology.

Thus we have

$$\vdash_T (\phi_A(\mathbf{n}, y) \to y = \mathbf{1})$$

Hence, using the UI Rule,

$$\vdash_T \forall y\, (\phi_A(\mathbf{n}, y) \to y = \mathbf{1})$$

that is, since $k = 1$,

$$\vdash_T \forall y\, (\phi_A(\mathbf{n}, y) \to y = \mathbf{k})$$

and so the second condition for representability is satisfied by ϕ_A when $n \in A$.

The case where $n \notin A$ and $k = 0$ is similar.

So both conditions for representability are satisfied by ϕ_A in both cases. Hence ϕ_A represents χ_A in T.

Solution 1.2

In using the Tautology Rule to deduce (1.8) from (1.5) and (1.7), we used the fact that $\neg\phi$ is a tautological consequence of $(\phi \leftrightarrow \neg\psi)$ and ψ, where ϕ is the formula G and ψ is $\theta_T(\ulcorner G \urcorner)$. For this to be a correct use of the Tautology Rule we need to check that

$$(((\phi \leftrightarrow \neg\psi) \mathbin{\&} \psi) \to \neg\phi)$$

is a tautology, which is easily seen from its truth table.

$(($	$(\phi$	$\leftrightarrow$	$\neg$	$\psi)$	$\&$	$\psi)$	$\to$	$\neg$	$\phi)$
	1	0	0	1	0	1	1	0	1
	1	1	1	0	0	0	1	0	1
	0	1	0	1	1	1	1	1	0
	0	0	1	0	0	0	1	1	0

In using the Tautology Rule to deduce (1.11) from (1.5) and (1.10), we used the fact that ϕ is a tautological consequence of $(\phi \leftrightarrow \neg\psi)$ and $\neg\psi$. To check that this is a correct use of the Tautology Rule we need to check that

$$(((\phi \leftrightarrow \neg\psi) \& \neg\psi) \rightarrow \phi)$$

is a tautology, which is easily seen from its truth table.

$(((\phi$	$\leftrightarrow$	$\neg$	$\psi)$	$\&$	$\neg$	$\psi)$	$\rightarrow$	$\phi)$
1	0	0	1	0	0	1	1	1
1	1	1	0	1	1	0	1	1
0	1	0	1	0	0	1	1	0
0	0	1	0	0	1	0	1	0

Solution 2.1

One suitable recursive function f for enumerating the set is given by

$$f(n) = \begin{cases} 2, & \text{if } n = 2, \\ 3, & \text{if } n = 3, \\ 8, & \text{otherwise.} \end{cases}$$

This f gives the values

$$\{f(0), f(1), f(2), f(3), f(4), f(5), \ldots\} = \{8, 8, 2, 3, 8, 8, \ldots\} = \{2, 3, 8\}.$$

The function f is of the form

$$f(n) = \begin{cases} g_1(n), & \text{if } R_1(n), \\ g_2(n), & \text{if } R_2(n), \\ g_3(n), & \text{if } R_3(n), \end{cases}$$

where

$$\begin{aligned} R_1(n) &\iff n = 2 \\ R_2(n) &\iff n = 3 \\ R_3(n) &\iff \text{not } R_1(n) \text{ and not } R_2(n) \end{aligned}$$

and

$$g_1(n) = 2, \quad g_2(n) = 3, \quad g_3(n) = 8.$$

It is easy to see that the relations R_1, R_2 and R_3 are mutually exclusive and exhaustive.

The characteristic function of R_1 is given by

$$\chi_{R_1}(n) = \chi_{eq}(n, 2),$$

which is primitive recursive as it is obtained by substitution from the primitive recursive function χ_{eq} using constants. Hence the relation R_1 is primitive recursive.

Similarly the characteristic function of R_2 given by

$$\chi_{R_2}(n) = \chi_{eq}(n, 3),$$

is primitive recursive. Hence the relation R_2 is primitive recursive.

That the relation R_3 is primitive recursive follows from Problem 1.10 of *Unit 2*.

Each of the functions g_1, g_2 and g_3 is a constant function and is thus primitive recursive.

Therefore, by Theorem 1.5 of *Unit 2*, the function f is primitive recursive, and hence total recursive, so that the set $\{2, 3, 8\}$ is recursive enumerable.

Solution 2.2

The empty set is recursively enumerable by definition. So we need only consider the case of a finite non-empty set, say $A = \{n_1, n_2, \ldots, n_k\}$. One suitable function f is given by

$$
f(n) = \begin{cases}
n_1, & \text{if } n = n_1, \\
n_2, & \text{if } n = n_2, \\
\vdots & \quad \vdots \\
n_{k-1}, & \text{if } n = n_{k-1}, \\
n_k, & \text{otherwise.}
\end{cases}
$$

The function f can be shown to be primitive recursive, and hence total recursive, along similar lines to Solution 2.1 and clearly $A = \{f(0), f(1), f(2), \ldots\}$. Hence A is recursively enumerable.

Solution 2.3

Suppose A is a recursive set of natural numbers. If A is empty, then A is recursively enumerable by definition. So we can suppose that A is not empty. Let n_0 be the smallest number in A. The characteristic function χ_A of A is recursive and total. Now let f be the total function defined by

$$
f(n) = \begin{cases}
n, & \text{if } n \in A, \\
n_0, & \text{if } n \notin A.
\end{cases}
$$

Then

$$
f(n) = n\chi_A(n) + n_0\,\overline{\text{sg}}(\chi_A(n)),
$$

so that f is obtained by substitution from the recursive functions χ_A, add, mult and $\overline{\text{sg}}$ using constants. Thus f is a recursive function. Also $A = \{f(0), f(1), f(2), \ldots\}$. Hence A is recursively enumerable.

Solution 3.1

Let $\phi(x)$ be the formula $(\mathbf{0} \cdot x) = \mathbf{0}$. We shall follow the strategy of Example 3.2. Consider the following formal proof.

1	(1)	$\forall x\,(x + \mathbf{0}) = x$	Ass
2	(2)	$\forall x\,(x \cdot \mathbf{0}) = \mathbf{0}$	Ass
3	(3)	$\forall x\,\forall y\,(x \cdot y') = ((x \cdot y) + x)$	Ass
4	(4)	$\phi(x)$	Ass
2	(5)	$\phi(\mathbf{0})$	UE, 2
3	(6)	$\forall y\,(\mathbf{0} \cdot y') = ((\mathbf{0} \cdot y) + \mathbf{0})$	UE, 3
3	(7)	$(\mathbf{0} \cdot x') = ((\mathbf{0} \cdot x) + \mathbf{0})$	UE, 6
1	(8)	$((\mathbf{0} \cdot x) + \mathbf{0}) = (\mathbf{0} \cdot x)$	UE, 1
1, 3	(9)	$(\mathbf{0} \cdot x') = (\mathbf{0} \cdot x)$	Sub, 7, 8
1, 3, 4	(10)	$\phi(x')$	Sub, 4, 9
1, 3	(11)	$(\phi(x) \rightarrow \phi(x'))$	CP, 10
1, 3	(12)	$\forall x\,(\phi(x) \rightarrow \phi(x'))$	UI, 11
13	(13)	$((\phi(\mathbf{0})\ \&\ \forall x\,(\phi(x) \rightarrow \phi(x'))) \rightarrow \forall x\,\phi(x))$	Ass
1, 2, 3, 13	(14)	$\forall x\,\phi(x)$	Taut, 5, 12, 13

Assumptions 1, 2 and 3 are axioms $Q4$, $Q6$ and $Q7$, and assumption 13 is an induction axiom. It therefore follows that $\vdash_{\text{PA}} \forall x\,\phi(x)$, that is $\vdash_{\text{PA}} \forall x\,(\mathbf{0} \cdot x) = \mathbf{0}$.

Solution 3.2

Following the hint we show that

$$\vdash_Q (\forall x\, (\mathbf{0} + x) = x \rightarrow \forall x\, (x \cdot \mathbf{0}') = x) \qquad\qquad (\text{S.1})$$

Consider the following formal proof.

1	(1)	$\forall x\, (\mathbf{0} + x) = x$	Ass
2	(2)	$\forall x\, (x \cdot \mathbf{0}) = \mathbf{0}$	Ass
3	(3)	$\forall x\, \forall y\, (x \cdot y') = ((x \cdot y) + x)$	Ass
3	(4)	$\forall y\, (x \cdot y') = ((x \cdot y) + x)$	UE, 3
3	(5)	$(x \cdot \mathbf{0}') = ((x \cdot \mathbf{0}) + x)$	UE, 4
2	(6)	$(x \cdot \mathbf{0}) = \mathbf{0}$	UE, 2
2, 3	(7)	$(x \cdot \mathbf{0}') = (\mathbf{0} + x)$	Sub, 5, 6
1	(8)	$(\mathbf{0} + x) = x$	UE, 1
1, 2, 3	(9)	$(x \cdot \mathbf{0}') = x$	Sub, 7, 8
1, 2, 3	(10)	$\forall x\, (x \cdot \mathbf{0}') = x$	UI, 9
2, 3	(11)	$(\forall x\, (\mathbf{0} + x) = x \rightarrow \forall x\, (x \cdot \mathbf{0}') = x)$	CP, 10

Since assumptions 2 and 3 are axioms $Q6$ and $Q7$, this formal proof shows that (S.1) holds. In Example 3.2 we showed that

$$\vdash_{\text{PA}} \forall x\, (\mathbf{0} + x) = x \qquad\qquad (\text{S.2})$$

Hence, applying the Tautology Rule to (S.1) and (S.2) shows that $\vdash_{\text{PA}} \forall x\, (x \cdot \mathbf{0}') = x$.

Solution 3.3

Let $\phi(x)$ be the formula $\forall y\, \forall z\, ((y + z) + x) = (y + (z + x))$.

We show first that $\vdash_Q \phi(\mathbf{0})$. Consider the following formal proof.

1	(1)	$\forall x\, (x + \mathbf{0}) = x$	Ass
1	(2)	$((y + z) + \mathbf{0}) = (y + z)$	UE, 1
1	(3)	$(z + \mathbf{0}) = z$	UE, 1
	(4)	$(z + \mathbf{0}) = (z + \mathbf{0})$	II
1	(5)	$z = (z + \mathbf{0})$	Sub, 3, 4
1	(6)	$((y + z) + \mathbf{0}) = (y + (z + \mathbf{0}))$	Sub, 2, 5
1	(7)	$\forall z\, ((y + z) + \mathbf{0}) = (y + (z + \mathbf{0}))$	UI, 6
1	(8)	$\forall y\, \forall z\, ((y + z) + \mathbf{0}) = (y + (z + \mathbf{0}))$	UI, 7

Since assumption 1 is axiom $Q4$, this shows that $\vdash_Q \phi(\mathbf{0})$.

Next we show that $\vdash_Q \forall x\,(\phi(x) \to \phi(x'))$. Consider the following formal proof.

1	(1)	$\forall x\,\forall y\,(x + y') = (x + y)'$	Ass
2	(2)	$\phi(x)$	Ass
2	(3)	$\forall z\,((y + z) + x) = (y + (z + x))$	UE, 2
2	(4)	$((y + z) + x) = (y + (z + x))$	UE, 3
1	(5)	$\forall y\,(w + y') = (w + y)'$	UE, 1
1	(6)	$(w + x') = (w + x)'$	UE, 5
1	(7)	$\forall w\,(w + x') = (w + x)'$	UI, 6
1	(8)	$((y + z) + x') = ((y + z) + x)'$	UE, 7
1, 2	(9)	$((y + z) + x') = (y + (z + x))'$	Sub, 4, 8
1	(10)	$\forall x\,\forall w\,(w + x') = (w + x)'$	UI, 7
1	(11)	$\forall w\,(w + (z + x)') = (w + (z + x))'$	UE, 10
1	(12)	$(y + (z + x)') = (y + (z + x))'$	UE, 11
	(13)	$(y + (z + x)') = (y + (z + x)')$	II
1	(14)	$(y + (z + x))' = (y + (z + x)')$	Sub, 12, 13
1, 2	(15)	$((y + z) + x') = (y + (z + x)')$	Sub, 9, 14
1	(16)	$(z + x') = (z + x)'$	UE, 7
	(17)	$(z + x') = (z + x')$	II
1	(18)	$(z + x)' = (z + x')$	Sub, 16, 17
1, 2	(19)	$((y + z) + x') = (y + (z + x'))$	Sub, 15, 18
1, 2	(20)	$\forall z\,((y + z) + x') = (y + (z + x'))$	UI, 19
1, 2	(21)	$\phi(x')$	UI, 20
1	(22)	$(\phi(x) \to \phi(x'))$	CP, 21
1	(23)	$\forall x\,(\phi(x) \to \phi(x'))$	UI, 22

Since assumption 1 is axiom $Q5$, it follows that $\vdash_Q \forall x\,(\phi(x) \to \phi(x'))$.

As the theory PA extends the theory Q, this means that

$$\vdash_{\text{PA}} \phi(\mathbf{0}) \quad \text{and} \quad \vdash_{\text{PA}} \forall x\,(\phi(x) \to \phi(x'))$$

Combining these proofs in PA with the induction axiom

$$((\phi(\mathbf{0}) \;\&\; \forall x\,(\phi(x) \to \phi(x'))) \to \forall x\,\phi(x))$$

enables us to deduce that $\vdash_{\text{PA}} \forall x\,\phi(x)$, that is

$$\vdash_{\text{PA}} \forall x\,\forall y\,\forall z\,((y + z) + x) = (y + (z + x))$$

INDEX

$GN(T)$ 7
PA   18
Prov(x) 23
$\delta(j)$ 13
$\phi(\mathbf{n})$ 17
$\phi(\tau)$ 17
$\Gamma(F)$ 13

axiomatizable
    recursively   11

Cantor   26
Church   27
Church's Theorem   10

enumerable
    recursively   11

Frege   27

Gödel   28
Gödel's First Incompleteness Theorem   15
Gödel's Second Incompleteness Theorem   25

Hilbert   28
Hilbert's Question   16, 22, 26

induction axioms   18

Kleene   28

Löb's Theorem   24
Leibniz   29
Leibniz's Question   4, 8

Mathematical Induction   17

Peano   29
Peano Arithmetic   18
Post   29
Principle of Mathematical Induction   17
provability predicate   23

recursive set of axioms   11
recursive set of theorems   11
recursively axiomatizable   11
recursively enumerable   11
representable set   5
Russell   30

Turing   30

101 CAKE DECORATING IDEAS

Patricia Simmons & Marie Sykes

NATIONAL
BOOK DISTRIBUTORS AND PUBLISHERS

ACKNOWLEDGEMENTS

Marie and Patricia would like to thank Ken Sykes for his encouragement and help in many practical ways during the preparation of this book.

Patricia would like to thank her life-long friends Beatrice Hailstone, for introducing her to cake decorating, and Veronica Bagley, for ever-present encouragement and confidence in her ability.

Published by National Book Distributors
3/2 Aquatic Drive, Frenchs Forest, NSW, 2086

First edition 1983
Reprinted 1984 (twice), 1986, 1988
Paperback edition 1990, 1991
Revised edition 1995
© Marie Sykes and Patricia Simmons 1983, 1995
Designed by John Bull
Typeset by Walter Deblaere and Associates
Printed in Singapore by Kyodo Printing Co (S'pore) Pte Ltd

Acknowledgements:
Australian Bakels Pty Ltd (Pettinice)
A. F. Bambach Pty Ltd (for wire)

National Library of Australia Cataloguing-in-Publication data

Simmons, Patricia.
    101 cake decorating ideas.
    New ed.
    Includes index.
    ISBN 1 86436 041 0.
    I. Cake decorating. I. Sykes, Marie. II. Title.
    III. Title: One hundred and one cake decorating ideas.

641.8653

101 CAKE DECORATING IDEAS

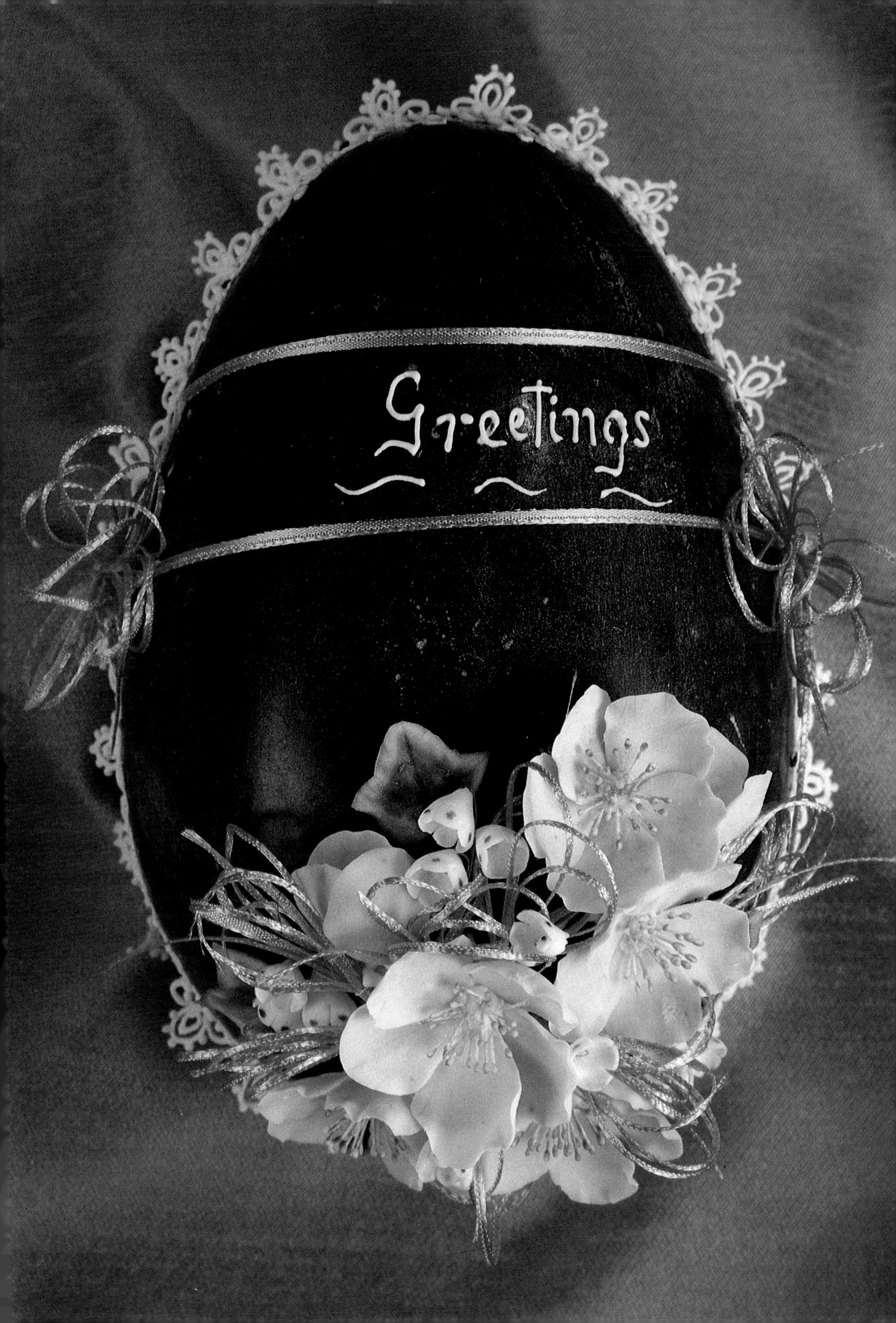
Greetings

CONTENTS

Introduction 7

I Equipment 8

II Recipes 10

III Supplementary Techniques 15

IV Moulded Flowers 21

V Wildflowers 29

VI Christening Cakes 35

VII Birthday Cakes 39

VIII Wedding Cakes 47

IX Anniversary Cakes 57

X Christmas Cakes 63

XI Novelty Cakes 69

XII Miscellaneous 79

Step-by-step Photographs 88

Patterns 97

Index 109

Introduction

This book is a companion to our first book, *Cake Decorating* and should be used with that book. This book supplements the information in the first book and also presents more than one hundred and one ideas for particular cakes.

Here, we give additional information on equipment, recipes and techniques, and two chapters of step-by-step instructions for making many more moulded flowers and wildflowers. Once again, we have chosen popular and unusual flowers which have not been fully illustrated elsewhere. The step-by-step photographs of the moulded flowers appear twice: once with the instructions and once in the large plates at the back of the book. These proved very popular and useful in our first book and we are sure they will again. The flowers have been moulded from nature as far as practicable.

In Chapters VI to XII of this book we have moved on from the technical material to give you ideas in full-colour photographs for more than one hundred and one cakes suitable for many celebratory occasions. Both beginners and more experienced decorators will find these a source of inspiration. Christenings, birthdays, weddings, anniversaries and Christmas are all given full coverage. A wide range of fun novelty cakes is also included.

As we have said before: for success, follow the basic rules and practise the techniques. Don't forget, practice makes perfect! Use our ideas (and any you may get from magazines, shows and exhibitions), but don't be afraid to experiment. Cake decorating is a very rewarding way to express your creative flair.

Warning: **Great care should be taken to ensure that cocktail sticks, wires, modelling clay, hobby dough and other non-edible items are used for display purposes only when used in sugarcraft. Particular attention should be paid to chalk when colouring. It is important to remember that non-toxic does not mean edible.**

I
EQUIPMENT

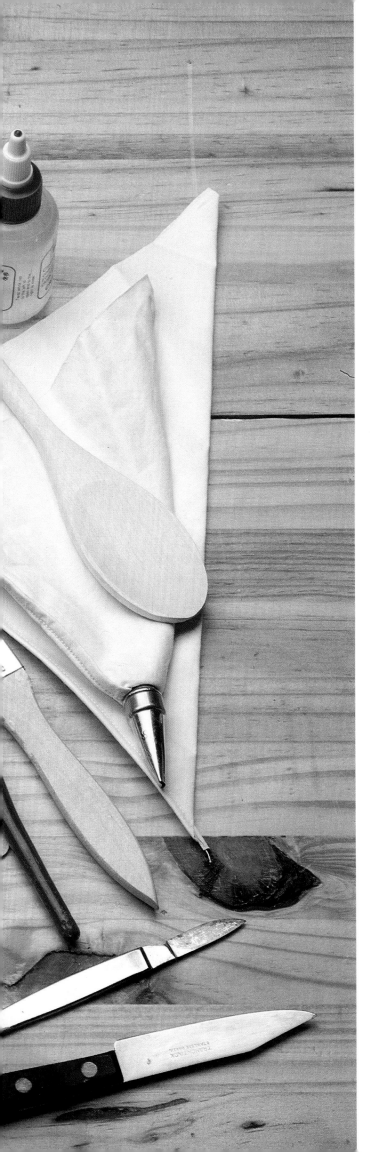

The equipment you will need for cake decorating is as follows.

Ten basic piping tubes and several screws: writing tubes 00, 0, 1, 2, 3; star or shell 5 and 8; petal 20 (small, medium or large) — left-handed people must use a left-handed petal tube; leaf 16; and basket 22. Additional tubes may be added later if required.

Paper bags made from good quality greaseproof paper.

Jaconette icing bags (small and large).

Petal and leaf cutters, for frangipani, fuchsia, etc., as well as for rose leaves.

Icing nails for piped flowers.

Scissors, fine-pointed for shaping small flowers, trimming leaves and petals, and a cheaper pair for cutting wire and paper, etc.

Tweezers with long points for arranging flowers and ribbons.

A very fine icing sieve (from health-food or specialist kitchenware stores).

Mixing bowls of glass or china.

Several new wooden spoons.

Commercial stamens in several shades.

A turntable of wood or metal.

Vegetable dyes — basic colours, red, sky blue, yellow, leaf-green, brown, scarlet, apricot, rose pink, mauve; other colours may be added as required.

Non-toxic pastels (a type of chalk) adds delicate highlights to colours.

Acetic acid helps royal icing to set firmly.

Gold and silver non-toxic enamel to highlight greetings and names on cakes.

Crimpers to press designs into covering fondant.

Other equipment includes a large piece of heavy plastic (to cover table or bench for cleanliness if necessary); fine plastic knitting needle; a small wooden (with pointed ends) modelling stick; plastic ball-end spike for modelling (the sort used to hold hair rollers in place); a piece of dowel rod about 20 cm (8 in) long for rolling modelling paste; long wooden rolling pin; wooden meat skewers (cut them in half and round the cut end with sandpaper); two or three good quality artist's brushes (sizes 00, 1, 2 or 3); maize cornflour; plastic airtight containers; waxed and greaseproof paper; a large hatpin; dressmaker's pins; adhesive tape; plastic ruler; cotton-covered wire (fine, medium, and heavy); sharp-pointed knives; penknife; craft knife or scalpel; round bottom patty tins; good quality paper paste; foam sponge cloths; foil; soft brush for dusting cake surface; two pastry brushes; moulds for Easter eggs; pillars for tiered cakes; and wooden skewers (from butchers). Health-food stores often stock .decorating equipment apart from obvious household items.

II
RECIPES

Royal Icing

Royal icing, a piping icing, should be made carefully, so take time to mix it well. Sieve icing sugar through a very fine sieve. Buy your pure icing sugar from a busy supermarket to ensure its freshness — squeeze or shake the box before purchasing as a double check. Allow time for the egg to come to room temperature. Then separate the white, making sure none of the yolk is included. Be sure all utensils and the wooden spoon are free from all grease, as this can ruin the icing.

1 egg white
¼ teaspoon liquid glucose (optional)
250 g (8 oz) pure icing sugar
2 or 3 drops acetic acid

Beat egg white and liquid glucose lightly in a glass bowl with a wooden spoon, add finely sieved icing sugar a tablespoon at a time, beating well between each addition until the mixture is thick and creamy. Add acetic acid and beat until blended. When the icing is ready for use it should form and hold a smooth peak when pulled away from the mixture. (Beating time is approximately 20 minutes.) Keep in an airtight container, or cover with a damp cloth or plastic film to prevent crusting.

Marzipan or Almond Paste

Marzipan, or almond paste, is used as an undercoating (a) to give a smooth finish to the cake; (b) to prevent oils and discoloration from the fruits seeping through; (c) to add extra flavour. It can be made from ground almonds, almond meal (a mixture of sweet and bitter almonds) or marzipan meal (ground peach kernels).

750 g (1½ lb) pure icing sugar
125 g (4 oz) ground almonds, almond meal
or marzipan meal
2 egg yolks
2-3 drops almond essence (optional)
2 tablespoons sherry or orange juice
2 egg yolks
4 teaspoons lemon juice
1 teaspoon glycerine

Sift icing sugar into a bowl, add almond meal and mix together. In a separate container place egg yolks, almond essence, sherry (or orange juice), lemon juice and glycerine. Blend well. Add to dry ingredients and mix to a smooth paste. Knead lightly, using a little sifted icing sugar. This quantity will cover a 250 g (½ lb) cake.

Rolled Fondant Covering

Rolled fondant covering is the finished coat to your cake.

A well-covered cake is the sign of a good decorator. With a good base to work on, each step becomes a challenge for perfection. A bad covering is hard to camouflage, so take time to do it well. Plan your day. Remember humid conditions are not good for handling fondant. Make the fondant a day or two earlier, remembering to sieve icing sugar thoroughly (two or three times if necessary). Finally when you are ready to commence covering the cake disconnect the telephone so there is no fear of an interruption.

ROLLED COVERING FONDANT No. 1

60 ml (¼ cup) water
1 tablespoon gelatine
3 tablespoons liquid glucose
3 teaspoons glycerine
Flavouring and colouring optional
7 cups (approx. 1 kg/2 lb) pure icing sugar

Place water in a double saucepan, add gelatine and stir over low heat until dissolved; *do not boil*. Remove from heat, add glucose and glycerine, stir until combined and allow to become cool, but not cold. Add flavouring if required. Place 5 cups sieved pure icing sugar in a basin, make a well in the centre. Add liquid and stir with a wooden spoon until the icing sugar is absorbed. Place in an airtight container and leave for 24 hours as this helps to eliminate air bubbles.

Knead balance of sieved pure icing sugar into the mixture until it is smooth and pliable and not sticky to touch. Add colouring with care at this stage. Place in a container or plastic bag and allow to stand for a further 1 hour. This quantity will cover a 2 kg (4 lb) cake.

ROLLED FONDANT No 2
(with egg whites)

1 kg (2 lb) pure icing sugar
125 g (4 oz) liquid glucose or light corn syrup
2 teaspoons glycerine
2 egg whites
colouring and flavouring (optional)

Sift icing sugar into a glass bowl, make well in the centre and add glucose (softened with hot water, if necessary), glycerine and lightly beaten egg whites. Mix with a wooden spoon, gradually bringing the icing sugar in from the sides of the bowl, until the mixture is a stiff paste. Turn out on to a board that has been dusted lightly with sieved icing sugar and knead until smooth. Add colouring and flavouring if required.

When it reaches the desired consistency the fondant should hold its shape when squeezed and it should leave no stickiness on the fingers.

SUPREME PLASTIC ROLLED FONDANT

A confectionery thermometer is required to make this excellent covering. Make it at least one week before it is required.

GROUP A
⅔ cup (5 fl oz) water
125 g (4 oz) liquid glucose
500 g (1 lb) crystal sugar
30 ml (1 fl oz) glycerine
1 teaspoon cream of tartar

GROUP B
⅔ cup (5 fl oz) water
30 g (1 oz) gelatine
125 g (4 oz) copha (or other solid white vegetable shortening)
2.25 kg (4½ lb) pure icing sugar (sifted)

Boil Group A to 120°C (240°F) or to soft-ball consistency when tested in cold water. Remove from heat and allow bubbles to subside. From Group B, dissolve the gelatine in the water, and add to mixture. Add chopped copha and allow to cool. Beat in half the pure icing sugar gradually. Place in a sealed plastic container and leave for not less than 1 week, or until required. Before using, knead in the balance of the sifted icing sugar until a plastic consistency is obtained. This mixture will cover a large two tier cake, a small three tier cake, or three single 250 g (½ lb) cakes.

APPLYING ROLLED FONDANT COVERING

Clean the surface on which you are to work, and dust it with pure icing sugar. Place fondant on the surface and use a long rolling pin that has been lightly dusted with pure icing sugar. Roll it out to about 10 mm (⅜ in) thick and large enough to cover the top and sides of the cake. Brush the cake lightly with beaten egg white; drape fondant over a rolling pin, then lift carefully on to the cake, making sure it is evenly distributed. Care must be taken not to stretch the fondant. Dust the palms of your hands with pure icing sugar. Smooth the top of the cake first to eliminate air bubbles; cupping your hand, work the corners. Use the palms of your hands to smooth the sides. With a knife held vertically, trim fondant into the base. Take care not to cut the fondant too short (patching is always visible).

Modelling Paste

Modelling paste is used for moulded flowers, leaves, ornaments, Christmas decorations, wedding bells, bowls, etc . Follow the recipe carefully as correct consistency is of prime importance, especially for fine petals.

30 ml (1 fl oz) cold water
2 teaspoons gelatine
1 teaspoon liquid glucose
160 g (5 oz) pure sifted icing sugar
Extra sifted icing sugar

Place water in double enamel saucepan, add gelatine and stir over low heat until dissolved *(do not boil)*; add glucose and stir until dissolved. Allow to cool, but not to become cold. Add 155 g (5 oz) sieved pure icing sugar, stirring until it is absorbed. Place in an airtight container and leave for 24 hours. The mixture should be firm and spongy when set.

When ready to commence modelling, take a small quantity of the mixture and knead extra sifted icing sugar into it, until it is of a similar consistency to plasticine, Modelling paste keeps for longer periods if this method is used.

Pastillage

Pastillage, a mixture of royal icing and powdered gum tragacanth, is used for modelling where strength is required, for instance, for houses, churches, etc. Pastillage sets firmly in most weather. Chemists stock powdered gum tragacanth, however it may have to be ordered in advance. Sprinkle half a teaspoon of powdered gum tragacanth into a cup of well-worked royal icing and beat thoroughly with a knife. Place in an airtight container, allow to stand for 24 hours. Take desired quantity from the container, and add enough pure icing sugar to form a pliable dough (like plasticine). Knead well. Do not store in the refrigerator. Store in a plastic bag in an airtight container.

Gum Paste

Gum paste is used for single shapes that can be modelled quickly, e.g., small animals on a log cake. It is not recommended for flowers.

500 g (1 lb) pure icing sugar
2 teaspoons gelatine
¼ cup boiling water

Sift icing sugar; thoroughly dissolve gelatine in boiling water, add to half the quantity of icing sugar in a large bowl. Knead well, adding more icing sugar until the mixture is no longer sticky. Keep in a plastic bag in an airtight container.

Warm Icing

250 g (8 oz) pure sifted icing sugar
good squeeze of lemon or orange juice
1 teaspoon butter

Place the cake on a prepared board or plate. Prepare a collar (a double thickness of waxed paper about 25 mm (1 in) wide); place around the top of the cake so paper is 6 mm (¼ in) above the top of the cake; secure with tape. Mix icing sugar and lemon juice to a firm consistency, add butter and place on low heat, stirring with a wooden spoon until the butter melts. While warm pour evenly over the top of the cake. Allow to stand until set.

Rich Fruit Cake

This quantity will fill a 200 mm (8 in) square tin; halve the mixture for 150 mm (6 in) tin; double it for 250 mm (10 in) tin.

250 g (8 oz) each currants, sultanas and raisins
90 g (3 oz) each dates and prunes
60 g (2 oz) each almonds, mixed peel, glacé cherries

(dried apricots, figs and pineapple, optional)
2 tablespoons each rum, brandy, sherry
250 g (8 oz) butter
250 g (8 oz) brown sugar
5 eggs
300 g (10 oz) plain flour
1 teaspoon each nutmeg, cinnamon, mixed spice

FLUIDS (all in one dish)

1 tablespoon plum jam
1 teaspoon vanilla essence
1 teaspoon Parisienne essence
2 tablespoons golden syrup
1 teaspoon glycerine
1 tablespoon lemon juice

Cut fruit and soak in alcohol. Leave for at least 24 hours. Cream butter and sugar, add eggs one at a time, beating well after each addition. Add fluids, then sifted dry ingredients and fruit alternately, mixing well. Place into a lined cake tin and smooth the top with a spatula that has been dipped in water. To release air pockets, drop the tin several times from a short distance. Place tin low in oven and bake at 120°C (250°F) for 3-3½ hours approximately. When cooked and while still hot, pour 2 extra tablespoons of sherry over the cake. Allow to cool in tin, then wrap in foil and a towel for six weeks to mature. (Baking times: 150 mm (6 in) 2-2½ hours; 250 mm (10 in) 6-6½ hours.)

Large Rich Butter Cake

This mixture will fill a 23 cm (9 in) square tin, 8 cm (3 in) deep. People ask on occasions for a butter mixture instead of the usual fruit cake; it is firm enough to hold a thin coating of marzipan and fondant covering.

375 g (¾ lb) butter
2 cups caster sugar
2 teaspoons vanilla
7 eggs
6 cups self-raising flour
1 cup (approx.) milk

Cream butter and sugar, add vanilla, eggs (one at a time) beating well after each addition so the mixture does not curdle. Add milk and flour alternately in small quantities, mixing thoroughly. Pour into a well-greased tin. Bake in a moderate oven for 1½ hours, until golden brown on top. Cool in the tin.

Genoise Cake

This recipe is suitable for making *petits fours*, or fancy shapes. Allow cake to stand for about 24 hours before cutting if possible, or place in freezer for a short time, then it will cut without crumbling.

60 g (2 oz) self-raising flour
1 tablespoon cornflour
3 eggs
125 g (4 oz) sugar
100 g (3 oz) butter or margarine, melted

Sift flour and cornflour. Place eggs and sugar in a basin over warm water and whisk lightly until the mixture is stiff enough to retain the impression of the whisk for a few seconds. Remove basin from heat. Sift half the flour mixture over the surface and fold in *very lightly*. Add balance of flour in the same way, followed by melted butter. Pour into greased and lined slab tin and bake in a moderate oven until golden brown (about 45 minutes depending on depth of tin).

Roses cakes may be made from this mixture. Cut shapes with a small scone cutter, coat sides with jam and toasted coconut or crushed nuts. Decorate the top with petals moulded from coloured marzipan in the shape of an open rose; for the centres decorate with chopped nuts or a glacé cherry.

Children's party cakes made from the genoise mixture may be baked in ice cream cones and decorated with hundreds and thousands, jelly beans or other small sweets.

Butter Icing

125 g (4 oz) butter
250 g (8 oz) pure sifted icing sugar
2 tablespoons sherry, lemon or orange juice

Cream butter and icing sugar, add liquid and beat until smooth. Spread evenly over the cake with a knife that has been dipped in milk.

Vienna Icing

Add 2 tablespoons sifted cocoa to the butter icing to make Vienna icing.

Marble Cake

This cake is used mainly for Dolly Varden cakes, its colours being one of the main attractions. A specially shaped Dolly Varden tin is used. This quantity fits into the standard sized tin.

250 g (½ lb) butter
250 g (½ lb) sugar
3 eggs
⅓ cup milk
vanilla essence
375 g (12 oz) self-raising flour
few drops red colouring
2 tablespoons sifted cocoa

Grease and line the tin. Cream butter and sugar, add well-beaten eggs, mixing well. Add milk and vanilla, then sifted flour. Divide the mixture in three equal parts: leave one part plain; add red colouring to the next; and add cocoa to the third. Place alternate spoonfuls in the tin. Bake in a moderate oven for 1 hour.

Flower Paste

3¾ metric cups pure sifted icing sugar
2 teaspoons (15 mls) gum tragacanth
2 teaspoons carboxy methyl cellulose (CMC)
2 teaspoons (10 mls) gelatine
2 teaspoons liquid glucose
3 level teaspoons Copha or white vegetable shortening

5 teaspoons (30 mls) cold water
¼ metric cup cornstarch
White of 1 large egg, with string removed

Soak the gelatine in 5 teaspoons of cold water until spongy, add glucose and copha or white vegetable shortening. Dissolve over hot water until clear.

Place the pure icing sugar and cornstarch into a baking dish, then sprinkle gum tragacanth and CMC over the top. Place in oven (200°) until warm to touch.

Place the dry ingredients into a warmed mixing bowl and with warmed beaters mix in the liquid ingredients on a slow speed until combined. Add egg white. Mix on a high speed until white and stringy (approximately 10 minutes depending on the mixer).

When mixed, roll in small pieces and wrap in plastic wrap to avoid crusting. Store in an airtight container for 24 hours before use.

When ready to use, take a small piece of paste, smear with white fat and add a spot of egg white, work until smooth and elastic. Allow to rest for a few minutes before use.

Practice makes perfect for fine flowers and if the first batch is not quite to your liking, try again.

Courtesy Denise Fryer (South Africa) and Tombi Peck

GUM GLAZE

1 teaspoon powdered gum Arabic
2 teaspoons hot water

Dissolve gum. The glaze is used for coating on marzipan fruits and leaves.

SUGAR SYRUP

¼ cup pure icing sugar
½ cup cold water

Dissolve sugar in cold water then bring slowly to the boil and simmer gently for 10 minutes; cool and bottle.
Store in the refrigerator.
Use as a stiffener for cotton net.

GUM GLUE

2 parts Pettinice commercial rolled fondant
1 part boiled water

Dissolve in microwave and cool. Repeat process.
This glue is used for attaching petals moulded in flower paste etc.

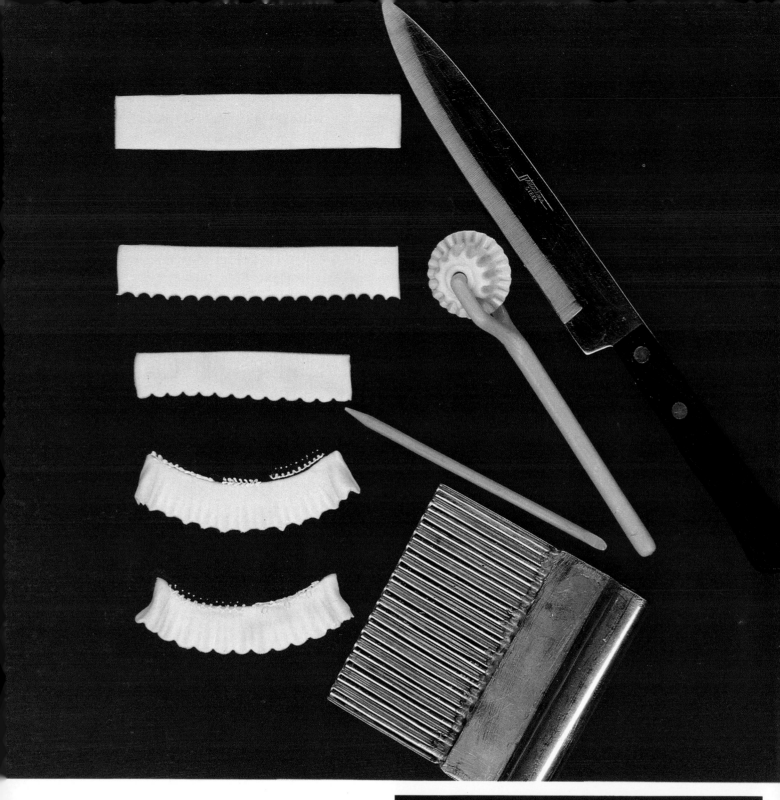

Above: Fondant frill and equipment
Right: Close-up of fondant frill

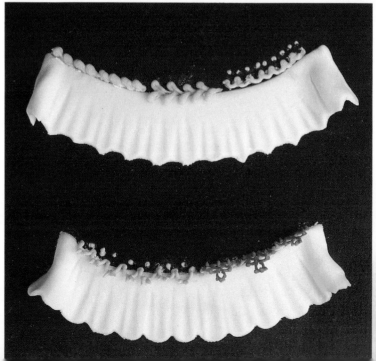

14

III
SUPPLEMENTARY TECHNIQUES

Pea Base

The pea base is a basic foundation used for all wildflowers of the Legume family such as broom, eggs and bacon, handsome wedge pea, Sturt's desert pea, kaka beak etc., and also for the sweet pea. Mould a small crescent-shaped bud and insert a hooked or knotted wire; prepare the desired quantity in advance and allow to dry.

Basic Hollow Cone

The basic hollow cone is the basis for many small flowers. Take a piece of modelling fondant the size of a small pea and mould to teardrop shape. Hold the shape between thumb, index finger and second finger and use a knitting needle, modelling stick, or small round toothpick, to hollow out the blunt end. Rest the cone on the index finger, at the same time rolling the knitting needle around the inside of the cone, concentrating on the edge.

Floodwork Variations

There are several techniques for floodwork once the selected pattern or design has been traced on to the cake or plaque using an HB pencil.

(a) A piped outline of the motif can be flooded with thinned-down royal icing.

(b) A slightly softened royal icing can be used without first piping a border. Small motifs are difficult to outline, but this method makes it possible. Use a very fine good quality paint brush and paint within the traced outline. In general work background detail first, major foreground detail last. Retouch small details when dry.

For example, a boat may be piped directly on to a cake and allowed to dry. Royal icing is then softened down with a little water to a thick spreading consistency and clouds are 'spread' with a small knife or spatula. Waves are formed in the same way, using a curling movement. Colour as desired when dry. If necessary, touch up edges with a very fine paint brush that has been dipped in brown colouring to accentuate the outline.

(c) Floral, geometric or motif designs may be piped on the sides of a cake and softened with a moistened brush to give a sculptured look. Butterfly wings on tulle or rice paper may also be done in this manner.

Rolled Fondant Frill

The rolled fondant frill is a very effective method of decorating the base of a cake. Edges may be trimmed with a pastry cutter or a scalloped cutter.

Step 1: Roll out strips of rolled fondant about 2.5 cm (1 in) wide and about 9 cm (3½) long.

Step 2: Trim one edge carefully with the scalloped cutter or pastry cutter.

Step 3: Using a small modelling stick, press it into each tiny scallop, flaring slightly.

Step 4: Repeat the process between each scallop to give a frilled effect. The imprint is made only two-thirds of the way up to the frill. This automatically gives the necessary fullness.

Step 5: Moisten the top of the frill with a little egg white or water and adhere to the cake.

The top edge may be finished in various ways: by piping a small shell edge, herringbone, a row of loops and dots, pipe-on lace or 'stand-up' lace.

Rolled Fondant Overlay

A rolled fondant overlay, another form of decoration, requires care in handling so that it is evenly distributed over the cake. Do not stretch. Cover the cake in the usual way and allow to stand for at least one day before proceeding to the overlay. Cut a pattern the size of the cake board.

Step 1: Roll out the rolled fondant and trim edges using a scalloped cutter or pastry wheel.

Step 2: Finger cut edges and flute slightly, using the small modelling stick.

Step 3: Brush the top of the cake lightly with egg white and carefully place the overlay in position.

Edges may be piped or trimmed as desired. The overlay can be slightly frilled with a plain edge and/or embroidered.

Eyelet Embroidery

Eyelet embroidery is very simple and quick, however it must be done neatly to be effective. Suitable designs are found in nursery motifs or handkerchief transfers, etc.

Select your design. Press the tip of a knitting needle into the rolled fondant. (The size of the knitting needle you use depends on the size of the hole required.) Using medium-peak royal icing and a fine tube pipe into the hole and then circle the rim finishing off neatly. (Eyelet embroidery tends to lift off at times, however, it is held firmly if this method is used. The same applies for any type of heavy embroidery.)

To pipe a neat circle, it is more effective to pipe half way round and then reverse the direction to completion. A pulled-dot from the centre and tiny dots around the rim are added variations.

Eyelet embroidery is suitable for all types of designs, particularly attractive when combined with the rolled fondant frill.

Embroidery

Embroidery is piped on the sides and sometimes on the tops of cakes to decorate a plain surface, and transform it to a work of art. Ideas for design can be found in many varied places — wrought iron, books, cards, papers, transfers and baby clothes. Other countries can provide beautiful designs, for example, Tyrolean or Spanish designs. Cross-stitch can also be used for a change.

Pipe with soft to medium-peak royal icing, incorporating light and heavy strokes for shading. This gives depth of character to a design.

If possible, do not pipe in humid or very hot weather as the surface is very soft and it is easy to mark with the tip of a tube. Get up early before the heat of the day, or work late in the evening if necessary.

Here we have illustrated roses, forget-me-nots, six petal and heart-shaped petalled flowers (blue background), lily of the valley (blue background), and hollyhocks, which have been incorporated into simple easy-to-follow designs.

When piping the cone of a hollyhock, first pierce the icing with a fine needle on the selected spot. Pipe a small amount of royal icing into the hole and continue circling the rim several times to form a cone. This method prevents the hollyhocks falling off.

Embroidery

Lily of the valley is made by first piping a curved stem. Pipe 3-4 small curves on the top side of the stem. Pipe four or five full rounded dots down the stem, graded from small to large. Allow a short drying time and then pipe three tiny pulled dots on the lower part of the full rounded dots to represent three small petals.

Extension or Dropped-String Work

Extension work (curtain border, bridgework or dropped string work) is a popular type of decoration for the base of a cake. It requires many hours of practice before perfection is attained. Beginners usually commence piping with No. 1 or No. 0 tube and, as experience is gained, graduate to finer tubes. In very recent years exhibition cakes have been piped with a hypodermic needle that has been soldered into a writing tube; the resultant thread is as fine as a cobweb.

However, beginners should concentrate at first on piping straight even lines before attempting to use finer tubes, and also on closing up the lines so there is no room to pipe another thread between them.

It is essential that royal icing be made with finely sieved pure icing sugar so that it runs freely. Very fine sieves are now available; silk, or a fine cloth may also be used.

Method

Step 1: Measure the circumference and height of the sides and cut a strip of greaseproof paper this exact size as a pattern. Fold the pattern in halves crosswise, then in quarters and so on until the desired size of a scallop is obtained. Cut a shallow curve through all thicknesses of the pattern to make the scallops. Shape the top of the pattern similarly, as desired. Place the pattern, scalloped side down, around the cake and secure ends of paper with adhesive tape, keeping the curves of the scallops close to the base. Mark the point of each scallop with a

Embroidery

pinprick on the rolled fondant. Then carefully mark the top edge of the design in the same way.

Step 2: Using No. 2 or 3 writing tube, pipe a shallow scallop on to the cake to cover the pinpricks; make sure the scallop adheres to the sides of the cake, otherwise the bridgework will collapse at a later date.

Step 3: Repeat the process of piping until at least four to five rows have been completed. Allow to dry completely before proceeding with the dropped thread work.

Step 4: Dropping threads. Use a fine tube (0, 00, 000) and soft peak icing. Work at eye level. Place the tube on the upper mark, and squeezing evenly, bring the tube out and down, and tuck it under the built-out work. The lines should be so close together that another line of icing cannot be piped in between any two lines. Neaten the bottom scallops with a tiny rope design, or dots, scallops, etc. Lace pieces may be added later to finish the top edge.

EXTENSION WORK VARIATIONS

Variations of extension work are shown here to give the decorator ideas of the different borders. Try some of your own incorporating some of our ideas.

(Left to right on the board.)

1. Split coloured ribbon is placed at regular intervals amongst the dropped thread with a band of the matching ribbon just above it. A fine snail's trail is piped below the ribbon and small dots neaten the base.

2. The split ribbon insertion has small dots piped on it. The top edge of the thread is piped with tiny scallops and the bottom edge is neatened with a single dropped loop. The piped-on lace covers the top of the extension work.

3. (left) Extension border with a double scallop, split ribbon and lace (as shown above) that has been piped on to waxed paper, allowed to dry and then placed in position with a little royal icing.

Extension work and lace

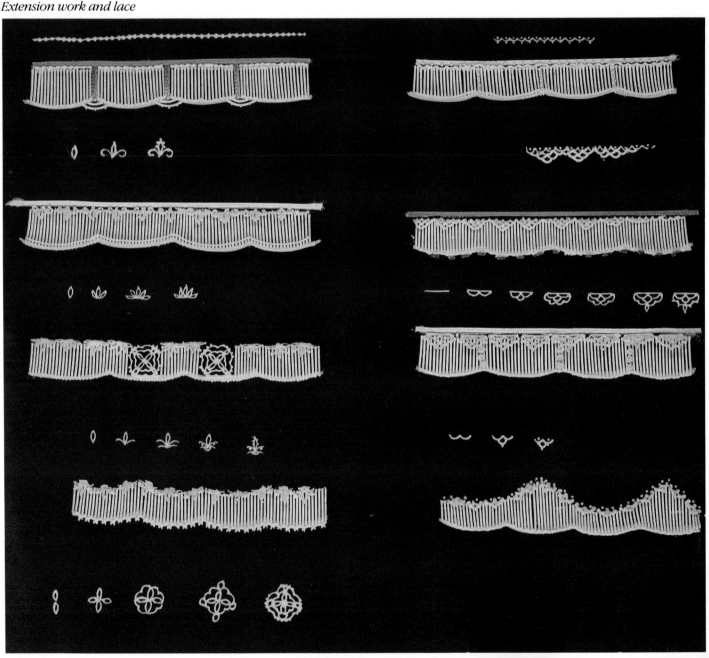

18

4. (right) Extension or dropped thread work features a ribbon band, piped-on lace which is finished off with tiny matching ribbon loops that have been attached with royal icing underneath the lower scallop.

5. (left) Extension or dropped thread work features lace medallion inserts. This type of work is suitable for advanced decorators only. Pipe the medallions and lace on to waxed paper and allow to dry. They must be joined continuously so that they may be transferred to the cake in one piece. The base is neatened with small dots.

6. (right) Pale blue forget-me-nots are piped on waxed paper, allowed to dry then transferred to the dropped threads, using a spot of royal icing to adhere them. The lace is piped on to waxed paper and then transferred when dry. Extension is neatened at the base with a scallop.

7. (left) Scalloped extension work requires a little more time. When the cake has been covered and allowed to stand for at least 24 hours to firm the icing, cut a paper pattern for the scallops.

  (a) Mark both top and bottom scallops with a pin.

  (b) Pipe small snail's trail with a fine tube at the base of the cake to hide the cut edges.

  (c) Change to No. 3 writing tube and pipe a row of scallops to cover the pin marks nearest the board.

  (d) Build out 3-4 rows, allowing drying time between each row. When the built-out work is completely dry, at least several days, pipe the dropped threads. When this is dry, add the lace at a 45° angle.

8. (right) The method for this variation is the same as above. The tiny lace is piped on and the extension work is neatened with a small scallop.

LACE PIECES

Lace pieces may be used to accentuate features or may be used with extension work. Suitable designs may be found on embroidery transfers, wallpapers, wrought-iron lace and lace trimming. Our designs will give you ideas, but try to create some originals of your own.

Paste half a sheet of graph paper on to a piece of masonite or heavy cardboard about 15 x 10 cm (6 x 4 in). Cover with waxed paper, firm and secure with adhesive tape. Mark out design.

Pipe uniformly and always pipe more pieces than required to allow for breakage. When thoroughly dry, carefully slide a small, rounded knife under the lace, loosening it from the waxed paper. Adhere to the desired spot with a little royal icing.

Pipe at least 100 pieces of lace for a 20-cm (8-in) square cake. Lace may be piped in advance and stored until required, taking care when loosening from the waxed paper.

PIPED-ON LACE

This is a very quick method for piping lace. It looks effective when piped with a fine tube.

1st row: Pipe a series of small loops covering the top of the dropped thread extension work.

2nd row: Pipe three small loops and miss one. Continue piping around the cake. This row stands out from the cake at a 45° angle.

3rd row: Pipe two small loops in the next round, making sure they stand out also as in the previous row.

4th row: Pipe one small loop to the end of the round. Finish with a dot on every single loop on the last round.

Variation:

1st row: As for the original 1st row.

2nd row: Pipe two loops, miss one.

3rd row: Pipe one loop, pipe a dot on the one loop to finish.

Subtle Colour in Extension Work

When the built-out frame work has been completed, dried, and before threads are piped, paint a strip of colour around the base of the cake, the depth of proposed dropped threads. Allow to dry. (Use a colourless alcohol mixed with the desired vegetable colouring.)

The colour shows through white or lighter coloured dropped threads. A matching ribbon will bring out a soft glow complementing the overall picture.

Bluebirds

Bluebirds

Bluebirds are used extensively on wedding and Christening cakes as they are a symbol of happiness. They are very dainty if piped with a fine tube. However, practice is required if they are to be perfected. The bluebirds are piped on to waxed paper, first the wings and then a tail, allowed to dry then assembled into a wet body.

Wings: Using medium peak royal icing and a fine tube, commence by piping a curve to the left and retracing, repeating the procedure three times, with each curve getting a little shorter. Pipe a curve to the right and complete the procedure, making a pair.

Tail: Commence piping at the point and retrace, pipe a smaller stroke in the centre, then another stroke to the right and overpipe again. The tail is similar to a V shape. Allow to dry.

Body: To make the body, place the tip of the tube on to the waxed paper, squeeze hard and move the tube slowly a short distance. Squeeze a smaller bulb for the head, relax pressure and pull away sharply for the beak. To assemble, set the tail in place. Set the dry wings into the wet body with the feathers pointing towards the tail.

Bluebirds may also be piped directly on to the cake using the dry wings and tail. As they are very fragile, care must be taken when assembling.

They may also be incorporated into embroidery on sides of cakes. First embroider your design, then pipe a wing and a tail flat on to the cake. Squeeze a bulb for the body, a smaller bulb for the head and pull off sharply for the beak. Place a dry wing into the wet body.

Colourings for Flowers, Plaques, etc.

Colourings may be used in a variety of ways:

(a) Vegetable colourings painted straight from the bottle for vibrant colours, or diluted with water, are the most widely used.

(b) Non-toxic chalks (applied with a brush as a powder) give a glow and highlight to petals, etc.

(c) Alcohol with a touch of vegetable colouring dries quickly on small flowers, eliminates the watermark and gives a soft hint of colour to the tips.

(d) A small atomiser filled with alcohol and a spot of vegetable colouring can be used to give a delicate realistic flower. Cover the centre of the flower with a thimble or small plastic top. Spray, allow to dry, and retouch the centre with powdered chalk.

Vary the sizes of flowers, leaves and buds, etc., for a realistic look.

Use an emery board to smooth rough edges of plaques, or if you have a broken petal, smooth it carefully in the same way, joining with craft glue.

The Flag Method

This is an easy way for making the centre of many roses.

Step 1: Take a ball of paste about the size of a 10-cent piece, bring the top to a point and press a waist into the centre, leaving the balance to fan out and form a stand. Place the stand on a firm surface. The top forms the bud; its size determines the size of the finished flower.

Step 2: Dust the forefinger and thumb lightly with cornflour and gently squeeze from the pointed top, down one-third of the bud, to form a small flag. Using a paintbrush moistened with water or egg white, brush the flag and wrap it around the tip to give a spiral effect.

This is now a tight bud, the centre of a rose.

Posy Arranging

Arranging posies is a very important part of cake decorating. Neatness is essential and flowers have to be presented at their best, joins and wires covered or concealed by ribbon loops and tulle fans. Small flowers and a few buds add lightness to the spray and green leaves add colour contrast.

Flowers may be wired if desired, using a 5 petal calyx.

Should the posy need elevation place a small ball of rolled fondant (about the size of a small walnut) on the desired spot. Do not use a large ball as it will need to be concealed by a large number of flowers. A small banana-shaped piece of fondant may be used for a crescent spray and a small carrot-shaped piece for longer spray.

MAKING THE POSY

Place all the flowers, leaves, etc., on a large piece of plastic sheet foam to avoid breakage. First place them into position on the plastic foam to see how they look, then proceed as follows:

If elevation is required secure a small ball of rolled fondant with royal icing on the desired spot. Using fine pointed tweezers, insert the wired ribbon loops around the small ball. Place flowers into position, adding tulle fans if necessary. Several ribbon loops may be required at this stage. Add small flowers and buds and finally leaves.

Use of Ferns in Floral Sprays

Preserved fern is frequently used in floral sprays to give a natural appearance. Numerous varieties are available — air fern, bracken, maidenhair and so on (both dried and fresh). Small sprigs only are used to achieve the desired effect. This method of using the real stem eliminates wire.

MAIDENHAIR

Step 1: Using fresh maidenhair fern, carefully remove leaves from the stems and place stem on waxed paper.

Step 2: Pipe a series of pale green graded dots on each stem in place of the fresh leaf.

Step 3: Allow to stand for a few seconds, place a small square of waxed paper over each frond and gently press flat. Leave the paper in position until dry and ready for use.

Step 4: Carefully peel off paper, retouch with a damp brush if necessary.

BABY'S BREATH

Baby's breath added to sprays, vases and posies on cakes gives a very delicate and airy effect. Spray lightly with hair spray before use to prevent the flowers from falling.

IV
MOULDED
FLOWERS

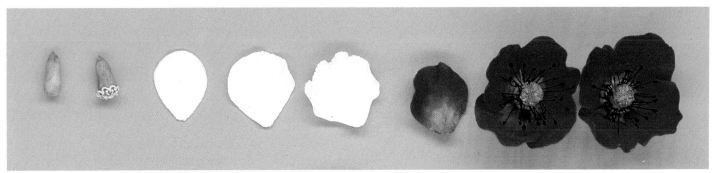

Corn poppy

Field Flowers

A traditional field flower posy consists of corn poppy, buttercup, field daisy, wheat and cornflower. They are a bright and colourful suggestion for the experienced decorator who likes 'something different' and are shown to advantage on a neutral coloured background. They may be used as the feature flower or combined with others. Some petal patterns and leaf patterns are given on p. 98.

Corn Poppy

The corn poppy is a spectacular deep red colour with a touch of black or green at the base.

Step 1: Mould a solid green gumnut shape pistil centre with a small tail; allow to dry.

Step 2: With royal icing, pipe eight tiny scallops around the rim, then a snail's trail from the centre into each scallop.

Step 3: Paint the piping on the top of the pistil black or green/brown and allow to dry.

Step 4: Cut four large petals; finger the cut edges, stretch, flute and vein. Cup base of petals slightly and allow to dry.

Step 5: Paint petals (i) blood red, or (ii) red with a touch of green at base, or (iii) red with a touch of black at the base. Allow to dry.

Step 6: To assemble, take a 5-cm (2-in) square of foil or waxed paper. Using firm-peak icing, No. 5 star tube and matching coloured icing, pipe a star in the centre of the square. Place the four petals in position, insert prepared

centre and surround with numerous black stamens. This method may also be used for the pale blue Himalaya poppies and the yellow Californian poppy.

Buttercup

Buttercups are part of the field flower selection. They are bright and colourful.

Step 1: Mould a dome-shaped centre with a short tail (6 mm) (¼ in) diameter. Allow to dry.

Step 2: Pipe a series of green pulled-dots over the centre of the dome. When dry paint with food colouring — a green-brown shade. Allow to dry.

Step 3: Cut five petals, finger the cut edges, cup petals, but do not flute. Allow to dry.

Step 4: Paint each petal a buttercup yellow shade and allow to dry.

Step 5: To assemble, using firm-peak royal icing in matching yellow, No. 5 tube, squeeze a small star of icing in the centre of a 4-cm (2-in) square of foil or waxed paper and arrange petals in spiral fashion, barely overlapping.

Place the moulded dome in the centre and fill in liberally with longer yellow stamens until the royal icing is covered.

Buttercup

Wheat

Step 1: To make main stem, use medium wire and mould a tiny bud in yellow to resemble a grain of wheat.

Step 2: Repeat the process. Mould wheat grains around the tip of the stamen. Approximately 24-36 are required for a stalk of wheat. Allow to dry.

Step 3: Tint small lengths of sisal or stamen cottons in a matching yellow. Allow to dry.

Step 4: To assemble, take the budded main stem, arrange three pieces of wheat a little way down from the first bud, winding matching yellow sewing cotton firmly down the main stem. Add about three wheat grains at a time interspersed with sisal or stamen cottons until the head is the required length. Do not crowd the grains. Seal the base with craft glue. It is not necessary to make the stems too long as only the head will be visible in the arrangement.

Wheat

Field daisy

Field Daisy

Step 1: Mould a small yellow dome-shaped centre on fine wire, allow to dry.

Step 2: Shape a basic hollow cone. Cut rim in halves, then cut each half into five equal parts.

Step 3: Working quickly, mitre each point halfway down the petal to make a long slender petal.

Step 4: With a knitting needle or a toothpick press each petal back gently against the forefinger which will automatically groove the petal.

Step 5: If necessary press each tip to a gentle point.

Step 6: Paint a spot of water at the base of the petals and pull the wired centre through gently. Press firmly to adhere and cut off any surplus. Keep the back as flat as possible.

Cornflower

Cornflowers bloom in numerous shades of blues, pinks (pale to dusty) and white. They are very attractive when moulded with matching stamens. To obtain a luminous glow add a touch of mauve or violet colouring to royal blue. Colour flowers in various shades of the selected colour for a more interesting arrangement.

Step 1: Tint modelling paste pale blue. Prepare a bud-base on medium wire. Allow to dry.

Step 2: Roll out a narrow strip of modelling paste about 8 cm (3 in) long. Using pinking shears double serrate one edge to make a fine fringe. Cut the strip into three or four pieces. (While working with one piece cover the remainder to prevent drying.)

Step 3: Stretch and vein along the serrated edge as finely as possible, gather the base to allow the serrated edge to represent three florets.

Step 4: While the florets are soft, arrange them around the prepared base, adhering with a spot of water to the dry centre. Add matching stamens. Approximately nine florets should fit around the base. Allow to dry.

Cornflower

Solomon's Seal

The green and white bells of Solomon's seal hang in clusters of twos and threes on a long arching stem. They may be used as an alternative to snowflake.

Step 1: Prepare fine wire, tint green, and make a small hook or knot on the end to prevent the flower from slipping off.

Step 2: Mould a long basic cone, divide rim into six equal sections.

Step 3: Mitre each to a shallow point and turn the petals inwards.

Step 4: Insert prepared wire into the centre of the flower.

Step 5: Tip petals with pale green colouring, some flowers have a green stripe. Arrange in clusters.

Solomon's seal   *Lilac*

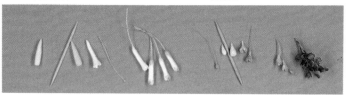

Lilac

Lilac grows in white, shades of pale mauve to deep purple, reddish purple, pink and soft yellow. There can be several shades in the one spray.

Step 1: Mould a basic hollow cone from a tiny piece of modelling paste. Cut rim into four equal parts.

Step 2: Trim corners and using the blunt end of the ball-end spike, shape and curve petals. Insert wire (use the finest gauge available). Allow to dry.

Step 3: Tint if necessary.

Step 4: Group three to five flowers with one bud for a spray.

White bauhinia

White Bauhinia

Bauhinia, a tropical flowering tree, is known as the orchid tree. Its blooms are white or pink, mauve or red. The white flower as illustrated is delicately tinted with pale green. Extreme care must be taken when assembling, as they are easily broken.

Step 1: Prepare five long brown-tipped stamens. Tint cottons green. Allow to dry. Prepare the stigma using a white stamen with a small narrow piece of modelling paste wrapped around the middle to resemble a miniature bean (seed box). Curve and allow to dry. Tint pale green.

Step 2: Cut two wing petals (see pattern, p.97). Vein and flute petals 1 and 2 and ruffle edges. Press over forefinger. Use a fine knitting needle to vein each petal. Place in a patty tin in clockface positions of 10 to 2 (petals turn back).

Step 3: Cut two lower petals (3 and 4). Repeat the process. Place in patty tin to dry in the 4 and 8 positions.

Step 4: Cut one throat, flute, vein and ruffle around the sides of the lip. Curve at the base and lay in patty tin with the lip overhanging. When dry paint the base of the throat a pale green colour. Dry.

To assemble, using very firm royal icing, No. 8 tube, pipe a star in the centre of a 5-cm (2-in) square of foil. Place petals 1 and 2 into the 10 and 2 clockface positions; petals 3 and 4 into the 8 and 4 positions, throat at 6 position, and prepared stigmas and stamens where 1 and 2 and base of throat meet. Allow to dry.

Pansy and viola

Pansies and Violas

Pansies and violas are brightly coloured in yellow, orange, brown purple and white. They may be tri-coloured with many colour combinations.

Step 1: Mould a five-petal green calyx, wire and allow to dry.

Step 2: Cut five petals (cover four while working on the one). Finger the cut edges, vein and flute lightly. Moisten back and place No. 1 petal almost upright on to the right side of the calyx. Give the petal a slight curl, do not allow to lie flat.

Step 3: Repeat the process for petal No. 2 and place it almost upright on the left side, slightly overlapping petal No. 1.

Step 4: Repeat the process and place petal No. 3 on the right side, slightly overlapping No. 1.

Step 5: Repeat the process and place petal No. 4 in the front of petal No. 2. Push petals into the calyx to secure.

Step 6: Vein and flute petal No. 5, push moistened pointed end of petal into the centre using the blunt end of the spike, groove the centre.

Step 7: Finish centre with a spot of royal icing; paint yellow or insert a large yellow stamen. Allow to dry and paint as desired.

When the flower is dry, bend the wire at the base of the calyx to a 90° angle for a natural appearance. Flowers may be assembled without calyx and wire if a flat arrangement is desired.

Geranium

Geranium

Geraniums bloom in pastel shades, bright terracotta shades and white. They are a good decoration for boys' cakes. Prepare a five-petal green calyx, add fine hooked wire and allow to dry. Colour the modelling paste calyx a basic colour and touch up when dry.

Step 1: Cut or mould five small petals, finger the cut edges. Do not flute.

Step 2: Set the five petals in spiral fashion. Secure them with a spot of water into the dry calyx. Petals should be slightly curved backwards. Make a small indentation in the centre of the flower using the blunt end of the spike.

Step 3: Add 3-5 stamen cottons, tip with a darker colour than the petals.

Step 4: Paint a few faint lines of a deeper shade on three lower petals.

Step 5: Mould tiny round buds in the same colour as the flower; paint a small calyx on the buds when dry. For a flower head take several flowers and lots of buds.

Ifala lily

Ifala Lily

The natural colour of the ifala lily is a deep rich salmon pink, however for decorating purposes they may be coloured as desired and used as a general small flower.

Step 1: Shape a basic hollow cone.

Step 2: Cut six shallow petals, mitre slightly. Press petals out gently and curve the bell.

Step 3: Insert fine hooked wire.

Step 4: Insert six matching stamens and one slightly longer stamen for the stigma.

Allow to dry and colour desired shade.

Easter daisy

Easter Daisy

Daisies are generally used in spring flower arrangements with daffodils, narcissuses, violets, etc. Although their natural colour is white with a soft yellow centre, they can be tinted to any desired shade for decorating purposes and used for most arrangements. Use fine, sharp embroidery scissors for the best results.

Step 1: Prepare a tiny yellow dome centre on fine wire and allow to dry.

Step 2: Take a piece of modelling paste the size of a small pea and mould to a basic hollow cone.

Step 3: Divide rim into four equal parts.

Step 4: Cut each section into as many petals as practicable.

Step 5: Moisten the centre and pull the wired dome through, keeping the back of the flower as flat as possible.

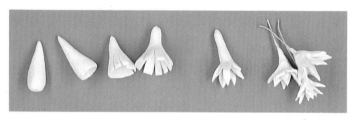

Tuber rose

Tuber Rose

Tuber rose is an off-white waxy type flower that grows on long spikes. It is very popular in wedding bouquets.

Step 1: Mould a large hollow cone, cut rim into 10 equal parts, mitre points.

Step 2: Moisten inside the cone, press sides together from base of cone to bottom of petal cuts, fold inwards from both ends simultaneously to form the centre, wrap remaining petals around to form the outer petals. Moisten lightly if necessary.

Step 3: If required on wire, push wire into base of the tube and firm on to the wire.

Narcissus

The narcissus family has a large variety of colours and shapes. Decorators have an infinite variety from which to choose.

Step 1: Mould a small ball of modelling paste into a cup using the ball-end spike or similar object. Insert fine-wire, and three short stamens. Allow to dry.

Narcissus

Step 2: Cut six small pointed petals, vein lengthwise across the finger to make the petals round.

Step 3: Adhere petals in two sets of triangles to the base of the cup. Remove surplus paste and allow to dry. Retouch if required.

Daffodil

Daffodil

Step 1: Mould a large cone, hollow, frill and flute the edge. (Leave slightly heavier at the base.) Insert wire if required. Add six yellow stamens and one stigma. Allow to dry.

Step 2: Cut six large, pointed petals, vein lengthwise and attach petals in two sets of triangles to the base of the trumpet. Remove surplus paste. Allow to dry.

Step 3: Retouch base colour if required.

Wood violet

Wood Violet

This particular violet, as its name suggests, grows in the woods or near hedges, and flowers in the spring. There are white and pink species, as well as the blueish-violet shades. They are very handy as an accent flower when a stronger contrast is required. Wood violets lend appeal to the spring flower collection. Prepare hooked wire, bending the hooked section to a 90° angle.

Step 1: Mould a basic hollow cone.

Step 2: Cut two narrow upper petals about one-quarter of the rim.

Step 3: Cut two wing petals about one-quarter of the rim.

Step 4: Leave the remaining paste for the base or throat petal.

Step 5: Mitre the upper petals, also the wing petals and trim the throat to resemble a round petal.

Step 6: With a small round modelling stick, press the two upper petals back against the finger and shape the two wing petals in the same way, but turned down.

Step 7: Slightly stretch the throat, lifting each corner in front of a wing petal. Then with the shaft of the modelling sticks ease the centre of the throat in an up-and-over movement to form the outstanding fifth petal.

Step 8: Insert moistened wire, making sure the angle of the wire is at the back of the flower (firm on to wire). This automatically forms the calyx. Add a deep yellow stamen. Allow to dry.

Paint calyx deep leaf green and remainder of flower the desired shade. Leave the base of each petal and the throat white.

Ground orchid

Ground Orchid

The ground orchid is a tiny flower that can be used with larger orchids or with a spray of small flowers. Prepare hooked wire, bending the hooked section to a 90° angle.

Step 1: Mould a basic hollow cone.

Step 2: Cut top sepal.

Step 3: Cut two wing petals.

Step 4: Cut two sepals, one below each wing petal.

Step 5: Leave the remainder for the throat.

Step 6: Mitre the three narrow sepals, and round the edges of the two petals and the throat.

Step 7: Using the modelling stick, press each sepal back against the forefinger.

Step 8: Stretch the two wing petals and throat. Shape the throat in the same way as for the wood violet. Bend the top sepal over slightly towards the throat. Insert prepared wire and finger back of flower as for wood violet. Colour as desired.

Iceberg rose

Iceberg Rose

This is a very handy sized rose to blend in with smaller or larger roses. Although the natural colour is white,

they can be shaded to suit any colour scheme. Use wire strong enough to support the weight of the flower.

Step 1: Take a piece of modelling paste, and using the flag method (p.20), make the tight bud.

Step 2: Mould three small petals each overlapping slightly; attach each to the bud with a spot of water or egg white.

Step 3: Mould five to seven petals to form the finished rose. Allow to dry.

Poinsettia

Poinsettia

Showy bracts of scarlet bright pink, or pale creamy-yellow form the spectacular part of this plant. They may be either assembled directly on to the cake or on to a square of foil and allowed to dry, then placed in position.

Step 1: Mould about 20 small flowers for the flower centre. For each flower take a tiny ball of green modelling paste and mould to teardrop shape. Allow to dry.

Step 2: Using royal icing, pipe an oval shape around the tip; several stamen cottons may be inserted.

Step 3: Using yellow royal icing, pipe a small oval shape on the side of the flower, then overpipe the cone. Allow to dry.

Step 4: Shape petal-shaped bracts by the freehand method. Finger the cut edges and vein. Run a toothpick down the centre for the vein and allow to dry on an irregular, but not flat, surface. Mould bracts in three sizes, make about six to eight of each. Shape some irregularly to create an interesting result. When dry, paint desired shade.

Step 5: To assemble: Using a No. 4 or 5 tube, squeeze a circle of firm-peak royal icing on to a square of foil or waxed paper. Place six to eight large bracts in an outer circle, then the middle-sized ones in the spaces and lastly the small bracts, taking care not to make the final result appear too flat. Squeeze another small circle of royal icing in the centre and add as many tiny flowers as practicable. Allow to dry.

Mould leaves.

Java Orchid

The Java orchid resembles a cymbidium, but is much smaller and very dainty. It is popular in bridal arrangements. The flower we describe has been patterned from nature — cream with pale green contrast tones. This orchid also grows in shades of pinks and greens.

Step 1: Mould a tiny column, attach to medium wire and allow to dry for 24 hours.

Step 2: Cut a throat (No. 6, see pattern) and finger the

cut edges. Place in the palm of the hand; roll a paintbrush handle from side to side to make the lobes stand up and curve inwards. Vein and stretch the lip and curve downwards.

Step 3: While still soft, attach to dried wired column. Allow to dry.

Step 4: Working quickly, cut or mould a pair of matching wing petals (Nos 2 and 3), finger the cut edges, vein, moisten and attach one on each side of the throat.

Step 5: Cut sepals (Nos 4 and 5), finger the cut edges, vein lightly, moisten and attach to underneath and on either side of the throat.

Step 6: Cut or mould sepal (No. 1), moisten, vein and attach to back of wing petals.

Step 7: Pipe two tiny lines on the base of the throat to resemble the pollinia. Allow to dry.

Step 8: Dust lightly across the backs of the lobes and throat with lime-green non-toxic chalk, or paint with food colouring. Tip column lightly with a fawn or brown colouring.

Java orchid

Waterlily

Waterlily

Waterlilies are found growing wild in the warmer parts of the world. They bloom in various colours including cream, pinks, yellow, red and violet.

Step 1: Colour modelling paste to desired shade. Mould approximately 21 petals, 9,7 and 5 of the large, medium and small respectively (see pattern, p. 98). Finger the cut edges. Roll a paint brush handle across the petal, pressing from left to right. Petals are curved lengthways, as well as crossways. Allow to dry.

Step 2: To assemble, take a square of foil or waxed paper, pale yellow firm-peak royal icing and pipe a small circle on the foil. Place large petals in a circle, then the medium sized ones, in alternate spaces, and lastly the tiny petals, overlapping the other two rows.

Step 3: Squeeze a star of royal icing in the centre and insert numerous yellow stamens. Allow to dry. Paint desired shade. Waterlilies can also be made using the frangipani petal cutters (small, medium and large), trimmed to a slender shape.

Browallia

Browallia is a popular flowering shrub. The buds are dark green and open to pale golden yellow. Then as the days pass they deepen to burnished shades of orange and deep copper. They are suitable for all types of cakes. Leaves are a deep contrasting green.

Step 1: Take a piece of modelling paste about the size of a pea.

Step 2: Mould a basic hollow cone, working out to a full flat circle, at the same time extend a long trumpet.

Step 3: Cut four even shallow petals and vein.

Step 4: Cut a tiny V in the centre of one petal; stretch and vein to resemble a scallop.

Step 5: Insert two yellow and one green-tipped stamens (showing only the heads).

Step 6: Allow to dry and paint as desired in varying shades of golden yellow, orange and deep copper. A tiny green calyx may be added if desired.

Step 7: Mould long and slender buds with a tiny calyx. Paint dark green when dry.

Browallia

Apricot or Quince Blossom

The tiny apricot or quince blossoms vary slightly in colour but the basic shape is the same. They are used in many arrangements and can be coloured as desired.

Step 1: Take a piece of modelling paste about the size of a small pea.

Step 2: Using a small modelling stick, mould a hollow basic cone.

Step 3: Cut rim into a five equal parts. Mitre corners.

Step 4: Vein and stretch each petal, cup with the ball end of the ball-end spike.

Step 5: Moisten centre and insert wire. Firm. Insert stamens while icing is soft.

Apricot or quince blossom

Jasmine

Jasmine is a five-petalled white flower with diluted soft burgundy undertones. Crescent-shaped buds are a highlight as their deep burgundy colouring stands out against the white. Leaves are small and dark green.

Jasmine

Step 1: Take a piece of modelling paste about the size of a pea, mould into a long teardrop shape.

Step 2: Form a basic hollow cone, cut rim into five equal parts.

Step 3: Trim corners and mould petals to a long and slender shape. Curl some back lightly.

Step 4: Insert prepared wire and finger a long trumpet. Allow to dry.

Step 5: Mould buds in a crescent shape, add wire.

When dry, paint the back of the flowers and trumpet with diluted burgundy colouring. Buds are painted a deeper shade.

Use plenty of buds when assembling and about 3-5 flowers.

Beloperone or Shrimp Plant

Beloperone is a mass of coloured heart-shaped bracts with a tiny white flower. The formation of these bracts resembles a hanging prawn. They are suitable for men's or boy's cakes.

Step 1: Cut a 6-cm (2½-in) piece of heavy wire and cover lightly with modelling paste. Allow to dry.

Step 2: Mould a tiny white flower with four long petals; add five brown-tipped stamens. Moisten and attach to dry base.

Step 3: Mould heart-shaped bracts and place in herringbone fashion, one overlapping the next. Turn the shape over and repeat the process. Beloperone has a four-sided appearance. Allow to dry.

Tint the bracts at the tip pale green, shading a pink-brown, leaving the small flower white. There is also a lime green and lemon variety.

Beloperone or shrimp plant

V
WILDFLOWERS

AUSTRALIAN FLOWERS

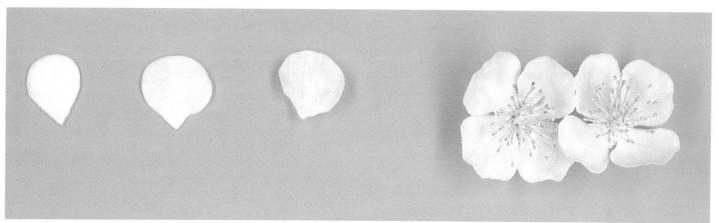

Leatherwood

Leatherwood

Leatherwood, a native shrub with leathery-type leaves, has dainty white four-petalled flowers and myriads of fine stamens.

Step 1: Stamp out four petals, finger the cut edges, stretch and vein.

Step 2: Place the petal in the palm and shape gently by applying light pressure with the cushion of the fourth finger. Mould three more petals.

Step 3: Place four petals in a shallow patty tin to dry.

Step 4: Assemble: Using firm-peak white royal icing and a No. 5 tube, squeeze a small star onto waxed paper or foil and arrange the four petals so that the points of the petals just meet in the centre. Squeeze another tiny star in the centre and fill with as many fine stamens as practicable. If fine stamens are not available, use stamen cottons tipped with a pastel colour.

Parakeelya

Parakeelya is a desert flower that blooms after heavy rains, colours vary from purple to white. Stamens are bright gold. Mould petals in cream modelling paste and paint the desired shade, leaving the base cream.

Step 1: Stamp out five petals.

Step 2: Stretch and vein, cup lightly.

Step 3: Allow petals to dry; paint desired shade.

Step 4: Assemble on a square of foil or waxed paper,

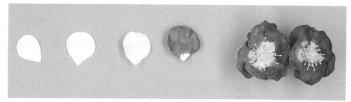

Parakeelya

using firm-peak royal icing. Arrange petals as illustrated: 1-2-3 in a triangle and 4-5 overlapping the three base petals.

Step 5: Squeeze another small star of royal icing, sprinkle liberally with fine gold stamens.

Halgania cyanea

Halgania cyanea

Halgania is a five-petalled blue flower with a striking centre in a dark contrasting colour. This flower blends well with the vibrant reds and yellows of other native flowers.

Step 1: Prepare the centre in a tiny bulb shape with stamen cotton protruding about 6 mm (¼ in) from the tip.

Step 2: Paint the base pale yellow and then paint five dark blue-brown stripes up to, and including, the stamen. Allow to dry.

Step 3: Colour modelling paste lavender blue, cut or mould five petals, each petal shaping to a gently rounded point. The petals are rather flat. Allow to dry.

Step 4: Touch up colour at base with chalk or liquid colour.

Step 5: To assemble, arrange in spiral fashion on a small square of foil or waxed paper using a minimum amount of firm-peak royal icing in a matching shade.

Step 6: Place centre in position securing with a small dot of royal icing. Allow to dry.

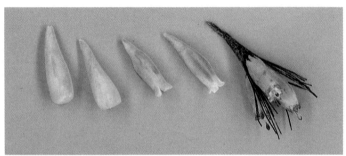

Pine heath

Pine Heath

Colour modelling paste yellow.

Step 1: Mould a long tubular shape, hollow the centre.

Step 2: Cut five shallow petals, mitre and curl tips back.

Step 3: Insert fine wire and one large stigma with dark brown head. Allow to dry.

Step 4: Paint tips a deep green.
  Leaves resemble pine needles.

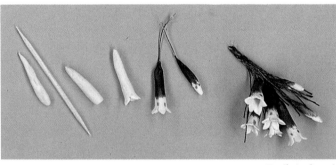

Fuchsia heath

Fuchsia Heath

Fuchsia heath is a native flower growing on a slender branch which blossoms in shades of pale to deep pink and reds. Mould in off-white modelling paste and paint as desired.

Step 1: Take a piece of modelling paste about the size of a small pea. Roll between the fingers to form a long tubular shape, approximately 2.5 cm (1 in) long.

Step 2: Hollow centre with a very fine knitting needle.

Step 3: Cut five small petals, shape with the fingers and curl back gently. Insert a fine prepared wire, firm at base. Allow to dry.

Step 4: Paint three-quarters of the flower with diluted burgundy or scarlet colouring, leaving the top petals white. Tip the base of petals with a spot of green. When dry paint a green calyx.

  Buds are tubular in shape. Wire and paint as for flowers.

Common correa

Common Correa

Common correa belong to a large family, they are long and tubular in shape and numerous in colour and colour combinations. They have four lobed petals, eight large stamens and one green stigma. They are something different to use in a wildflower arrangement.

Step 1: Take a piece of modelling paste about the size of a large almond. Hollow centre to resemble a tube and cut four shallow petals around the rim. Heavily vein each petal and turn petal back slightly.

Step 2: Insert hooked wire and firm; insert eight long stamens and one green stigma.

Step 3: Mould a cup-shaped calyx. Firm on to base of flower.

  Arrange in sprays of three to five flowers. Paint as desired.

Native fuchsia

Native fuchsia

Native fuchsia is a tubular showy flower, usually red, orange or plain yellow in colour. Some have a spotted lower lip. The flower has five curved stamens and one heavy stigma.

Step 1: Take a piece of modelling paste about the size of an almond, mould a long hollow cone and divide rim into five equal sections.

Step 2: Slash two adjacent fairly deep cuts for the long petal; the other three cuts should be about half the depth.

Step 3: Finger petal tips, round slightly, vein and turn the long petal back into a tight curve. The four small petals are curved back lightly. Insert prepared wire. Curl and insert five stamens and one heavier stamen for the stigma.

Allow to dry, then paint as desired. Paint calyx at base of flower.

Group in threes, plus two buds. Buds are almond shaped and may be wired. Paint calyxes when dry.

Wild violets

Wild Violets

Wild violets are small dainty white flowers with violet-coloured trim and a yellow centre. They grow in shady areas along bushland creeks and make a dainty addition to a spray of wildflowers.

Step 1: Take a piece of modelling paste about the size of a small pea.

Step 2: Make a basic hollow cone.

Step 3: Make two cuts (about one-third of the rim apart). Make a further two shallow cuts on the lower side of each of the first cuts, leaving the one larger piece for the lower petal.

Step 4: Mitre the two side petals, vein and stretch. The petals point to the side. Using the ball-end spike, cup the two small upper petals and also the larger lower petal. Shape the back of the flower to a point.

Step 5: Insert prepared fine wire (just above the back point). Firm wire. Tilt the flower downward to give a realistic appearance. Allow to dry. Leaving a rim of white around the edge, paint an irregular circle of violet. Tint the centre yellow, or add a yellow stamen (after inserting the wire).

Pine Cones

Take a piece of modelling paste about the size of a large pea. With the points of small sharp scissors, commencing at the top, snip around the fondant, moving in a circle, until the whole piece is covered with V-shaped serrations. Push each point so they stand out. When dry, paint any shade of brown. Arrange three to five in a group, depending on the size, with plenty of green stamen cottons to represent pine needles. Secure stamens with royal icing. Wire if required.

Pine cones

NEW ZEALAND FLOWERS

Puriri   *Pohutakawa*

New Zealand wildflowers are very colourful, unusual, and deserve a wider recognition by decorators. Their bright colours and unusual formation of petals enhance the most simple arrangement.

Puriri

The puriri tree has red or bright pink flowers and makes an attractive addition to the New Zealand wildflower collection.

Step 1: Take a ball of pink modelling paste about the size of a large pea.

Step 2: Hollow into a bell shape.

Step 3: Cut circumference into four shallow petals. Trim edges and cup, using the ball-end spike.

Step 4: Insert four white stamens with brown tips and a green or red stamen which is longer, for the stigma.

Step 5: Paint and allow to dry.

Step 6: Red berries may be added to the spray.

Pohutakawa

Pohutakawa (the New Zealand bottlebrush) blooms bright red. Select firm stamens so that they are easy to handle; cottons may be sprayed with sugar syrup if too soft.

Step 1: Colour stamen cottons red and allow to dry.

Step 2: Mould a shallow basic cone in pale green modelling paste, allowing the cone to remain thicker at the rim.

31

Step 3: Cut stamens to approximately 2.5 cm (1 in) in length. Using fine tweezers carefully insert as many stamens into the rim as possible, taking care not to damage the cone. Insert some fine yellow stamens as well as the red cottons.

When arranging add buds.

Manuka

Manuka

Manuka flowers are pink, white and red in colour, and range from very small flowers to those 2.5 cm (1 in) across.

Step 1: Mould a small piece of modelling paste to a hollow cone.

Step 2: Cut five even petals and mitre.

Step 3: Using the ball end of the spike, stretch and cup each petal. Hollow the centre of the blossom with the blunt end of the spike and insert wire.

Step 4: When dry, paint the hollow centre a soft green or delicate brown and pipe a series of pink pulled dots around the base of the petals for stamens.

Tint petals desired colour.

Chatham Island Forget-Me-Not

Chatham Island Forget-Me-Not

Chatham Island forget-me-not is a large forget-me-not. It has white-edged petals and is deep blue towards the centre of the flower.

Step 1: Take a piece of modelling paste about the size of a pea, mould into teardrop shape and hollow wide end to form a full circle.

Step 2: Cut rim into five equal shallow petals and mitre.

Step 3: Using the ball-end spike, cup and stretch petals.

Step 4: Insert medium hooked wire, firm on to wire and allow to dry.

Step 5: Paint, see illustration.

Kaka Beak

Mould a small pea foundation on wire. Allow to dry.

Step 1: Using the small frangipani petal cutter, stamp

Kaka beak

out a base petal, finger the cut edges, fold in halves lengthwise, moisten and press around the dry pea foundation, pointed end facing down to form a crescent.

Step 2: Stamp out a small petal, cut in halves lengthwise, finger each half and shape to a crescent. Moisten and attach these crescent-shaped wing petals (one on either side) at the base with tips upturned. Allow to dry.

Step 3: For the posterior petal, cut a similar size to the base petal; finger, vein crease gently and adhere to the base with a spot of water. Curl the petal back gently. Allow to dry.

Step 4: Paint red.

Step 5: Mould a cup-shaped pale green calyx. Allow to dry.

Buds are moulded in a crescent shape, in pale green, with a cup-shaped calyx.

Kowhai

Kowhai

Kowhai is one of the most beautiful and popular of all New Zealand flowers. Blooms are golden yellow and when the petals drop they leave a brown pod. Their leaves uncurl like fern fronds. Keep the flowers small for decorating purposes or they tend to look too heavy.

Mould a small pea base, insert wire and five stamens (as shown). Allow to dry.

Step 1: Cut three small petals. Cover two remaining while working on one. Vein and stretch petal, crease gently in halves lengthwise, moisten and attach to the pointed end of the dry pea foundation, leaving the stamens to protrude a little.

Step 2: Repeat as for Step 1, allow the next petal to fall away from the first petal.

Step 3: Cut the third petal in halves lengthwise and shape with the ball-end spike. Moisten pointed tip and attach to base of flower as illustrated.

Step 4: Mould a cup-shaped calyx. Allow to dry. Tint calyx burnished brown and the petals golden yellow.

Clematis

Clematis is a vine with beautiful white snowy blossoms and pale golden stamens. Maoris do not pick these flowers, as they see picking them as a bad omen.

Clematis

Step 1: Prepare centre: hook wire and mould a small pea-sized dome. Insert medium wire. While still soft insert at least 24 stamens. Allow to dry.

Step 2: Take a piece of modelling paste about the size of a small almond, shape to a basic hollow cone (see p. 15), cut the rim in half, and each half into four equal parts making deep cuts. Round the edges.

Step 3: While working quickly, press petals to a reasonable shape between the thumb and forefinger then lightly vein and stretch.

Step 4: Moisten base of cone, then pull the prepared centre into place. Firm on to wire and cut off any surplus at the back of the flower to keep it as flat as possible. Allow to dry. Give some petals a slight twist or lift. It is not a very flat flower.

Step 5: Touch up the centre with pale golden yellow to match the stamens.

Mountain lacebark

Mountain Lacebark

Mountain lacebark is a flowering tree that blooms in autumn. It has five petals and numerous long white stamens.

Step 1: Prepare the centre first. Hook wire, mould a small dome shape, insert medium wire. While still soft insert at least 24 long white stamens. Allow to dry.

Step 2: Take a piece of modelling paste about the size of a large pea. Shape into a basic hollow cone (see p. 15). Cut into five equal parts; the petals are long and slender.

Step 3: Press petals to a reasonable shape between the thumb and forefinger, lightly vein and stretch.

Step 4: Moisten base of cone, then pull the wired centre through. Firm on to the wire and cut off any excess at the back of the flower, keeping it as flat as possible. Give petals a curl as they are not rigid but fall in a starlike manner. Allow to dry.

New Zealand Eyebright

New Zealand eyebright is a white sub-alpine flower with vivid contrast colouring which makes an attractive addition to the collection.

New Zealand eyebright

Step 1: Take a piece of modelling paste about the size of a small almond. Hollow and shape a basic cone, divide rim into ⅓ and ⅔ sections.

Step 2: Shape the smaller piece to resemble 2 joined petals (heart shaped).

Step 3: Mould the larger piece to resemble 3 joined petals.

Step 4: Insert 5 stamens and 1 brown-tipped stigma.

Step 5: Vein heavily.

Step 6: When dry, touch up the veins with light green and the lower middle petal with a spot of golden orange. Paint back of the flower at the base of the 2 joined petals a golden orange. When dry, paint a brown-green calyx.

Mount Earnslaw ourisia

Mount Earnslaw Ourisia

Mount Earnslaw ourisia is a white sub-alpine flower.

Step 1: Take piece of modelling paste about the size of a small almond, hollow to a basic cone, divide rim into ⅓ and ⅔ sections.

Step 2: Cut the larger section into three equal parts. Round the edges, press the petals between the fingers, vein and stretch.

Step 3: Shape the smaller section to resemble 2 joined rounded petals (heart shaped).

Step 4: Moisten at the base and pull prepared wire through the centre, firm on to base. Add 5 stamen cottons and 1 white stamen for the stigma. Allow to dry.

Paint petals at the base a mixture of green and lemon and also paint the calyx dark green.

Poroporo

Poroporo is a shrub with a bell-like flower that blooms in dark purple, light blue and white.

Step 1: Prepare centre (a small rounded piece of modelling paste), insert five heavy yellow stamens, wire and allow to dry.

Step 2: Take a piece of modelling paste about the size of a small almond. Mould to a large hollow cone and cut into five equal shallow petals.

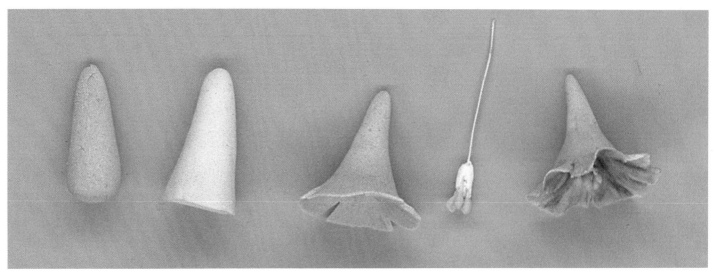

Poroporo

Step 3: Mitre, vein and stretch the circumference.

Step 4: Mark five centre veins using a toothpick.

Step 5: Insert prepared centre and allow to dry. Paint desired colour.

Kotukutuku or New Zealand Native Fuchsia

Native fuchsia — kotukutuku — is a small tree with silvery-backed leaves. Buds are green and purple-backed, flowers deep red. They also have an edible black-purple berry called konini.

Step 1: Group 5-6 stamens (including 1 longer yellow-tipped stamen) and bind together firmly with wire.

Step 2: Take a piece of modelling paste about the size of a large pea, model a basic hollow cone.

Step 3: Make 5 equal cuts (half-way down), mitre the ends, finger the cut edges, vein and curl petals back giving them a slight twist.

Step 4: Insert prepared stamens.

Step 5: Add a small ball to the base for the ovary, then a longer tubular shape (part of the stem) which is painted green.

Green buds are tubular shaped with purple-black streaks. Paint as desired and allow to dry.

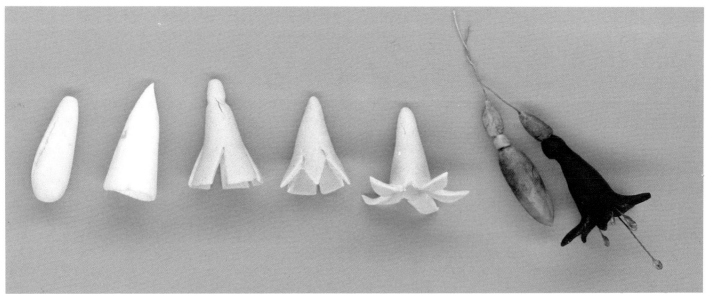

Kotukutuku

VI
CHRISTENING
CAKES

The feature of a Christening or name-giving ceremony
can be the decorated cake. Small nursery motifs, baby animals
with tiny flowers are popular. Moulded bibs
and bootees, when combined with flowers,
are very popular and attractive.

Crimper work and a fancy shell border with
ribbon insertion combine with a floodwork
plaque and delicate geraniums to make
this cake an eye-catcher.

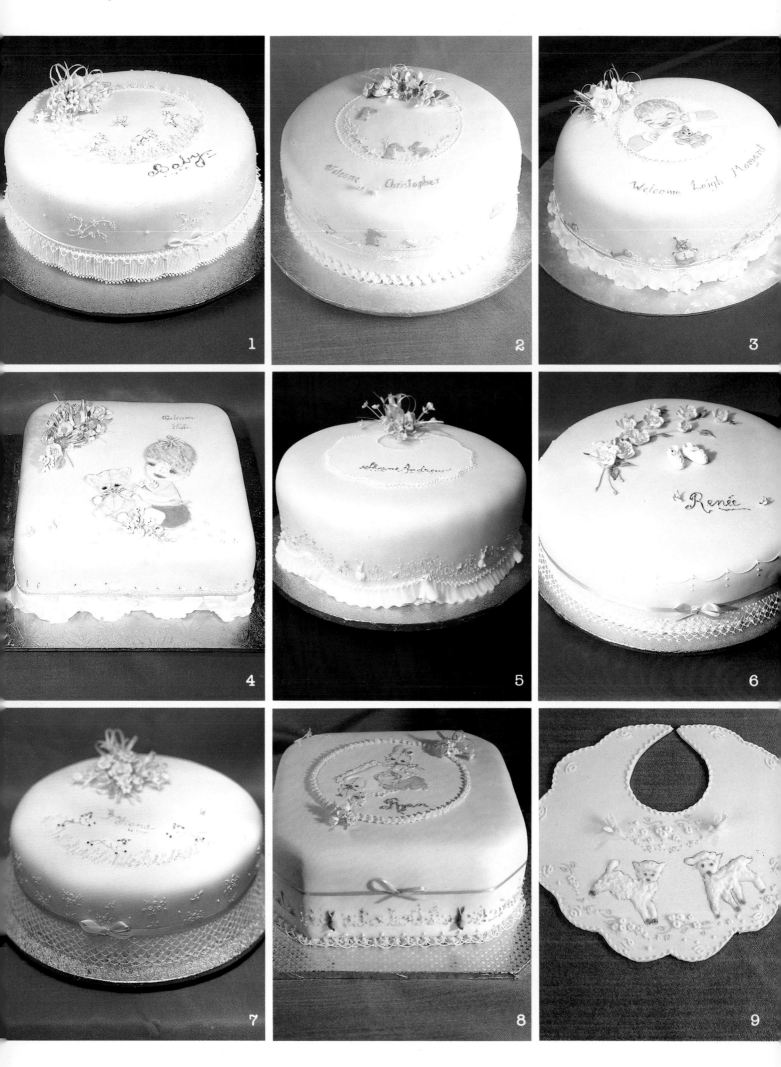

1. Flooded spring lambs and butterflies are the main feature of this cake. Ribbon is incorporated in the extension work. Embroidery is simple. A posy of flowers gives the finishing touch.

2. This cake is decorated with a posy of Cecil Brunner roses and wild violets. There is a simple piped base and decorations of flooded rabbits and chickens.

3. There are toys flooded onto the sides of this cake, simple embroidery and a double fondant frill; it also features a posy of small flowers.

4. This cake features a Christening motif of floodwork, piped bluebirds, embroidered swans and cygnets, a single fondant frill, with a posy of small flowers.

5. A cream bib with tiny loops is the main feature of this cake; a rolled fondant frill and coloured embroidery add the finishing touches.

6. Pink bootees and a spray of pale pink blossom and a mesh base edging are an attractive combination for a Christening cake.

7. Flooded lambs, eyelet embroidery and a posy of small flowers decorate this dainty cake.

8. A plaque flooded with nursery motifs, piped nursery designs on the side and a border of simple stars with alternate dropped loops is very suitable for a Christening cake.

9. This plaque features a pale pink bib with spring lambs and flowers.

10. A plaque suitable for twins, girl and boy, shows rabbits in field of flowers.

11. A stork delivering a baby girl surrounded with flowers is featured on this plaque.

10   **11**

Frangipani, forget-me-nots, hyacinths
and snowflakes make a very pretty spray.
Coloured embroidery and split satin ribbon
on the sides of the cake are the highlights.

VII
BIRTHDAY CAKES

Birthdays are celebrated from the first anniversary onwards
and frequently with a birthday cake.

The first cake is usually a sponge mixture decorated
in a very simple way. As the child grows older, butter cakes
are popular. For instance, the Humpty Dumpty, teepee,
the bride doll or doll's house in Chapter XII, suitable for
children's birthdays, all have a butter cake base.

Fruit cakes are used for the majority of birthday
celebrations. The firm cake base is suitable for marzipan and
rolled fondant coverings. Embroidery and extension work
are shown to advantage on the smooth fondant covering.
Flowers can be vivid in colour or a delicate pastel
—there are many from which to choose.

This simple cake includes full blown roses
and snowflake in a delicate posy, with
hollyhock embroidery and a simple border.

1. The posy includes apricot and quince blossoms, hyacinths and mini-hippeastrums. The cake also features a scattered embossed rose embroidery with a mesh border.

2. A posy of cream double and single fuchsias, hyacinths and snowdrops with lots of ribbon loops, plus hollyhock embroidery and a fine extension base decorate this cake.

3. A very simple vintage car plaque, cornelli decoration on the top of the cake and two small sprays of wildflowers — flannel flowers, wattle, brown boronia and a star-loop border — make this an attractive birthday cake for a boy or man.

4. This nautical cake is complete with anchors and lifebuoy, piped sailing boats on the sides and a posy containing approximately 36 daisies. The simple shell border is finished with loops and dots.

5. Full blown Peace and Cecil Brunner roses, hyacinths, snowflakes and gypsophila form the spray; eyelet embroidery and a rolled fondant frill complete the design.

6. Piped flowers in various shades combined with a moulded horseshoe, split ribbon and drop loop edging with a shell border is a popular cake.

7. Something different for a male who does not wish to have flowers on his cake—a bunch of keys made in modelling paste and painted, nandina (coloured leaves), cotoneaster berries and ivy leaves are ideal, with a very simple puff border.

8. A horseshoe shaped cake features moulded spring flowers — daffodils, jonquils, daisies and violets.

9. Formal roses, small Cecil Brunner roses, pink forget-me-nots, blossoms, snowflakes and fern are ever popular. Eyelet embroidery with a small matching posy on the base of the cake and a rolled fondant frill complete the decoration.

10. Single roses, violets, Easter daisy and gypsophila are a feature of this cake. A puff border and simple dropped loops on the sides complete the design.

11. Fresh jasmine with buds were taken from the garden so that the jasmine in the posy could be moulded from nature. Coloured embroidery on the sides and dropped star-loop border finish the cake.

12. Variation of formal roses using stamens in the centre, together with buds, numerous ribbon loops and leaves make up the posy. The crimper work is featured with embroidery and ribbon and a shell and star border finish the base.

13. The five posies which include Cecil Brunner roses and tiny mauve blossoms decorate this very large single cake, which has delicate side embroidery with fine extension work.

14. Pale pink blossoms and sweet peas decorate this tailored cake. Scattered embossed roses, ribbon, lace, and embroidery and extension work complete the decoration.

15. A sailing boat has been flooded directly on to the cake, and the royal icing 'roughed-up' to form the rough sea and clouds. A simple shell border with loops finishes the decoration.

16. This 'one-year-old' birthday cake for a girl features one candle and two fairies dressed in tiny piped skirts amongst a posy of blossoms, Cecil Brunner roses, hyacinths and snowflakes. The sides feature freehand coloured piping with a simple shell border with dropped loops.

17. The beautiful lady has been flooded directly on to the cake. Her posy of roses, violets, blossoms, daisies and maidenhair fern are moulded. Medallions of lace are combined with the fine extension work and lace pieces.

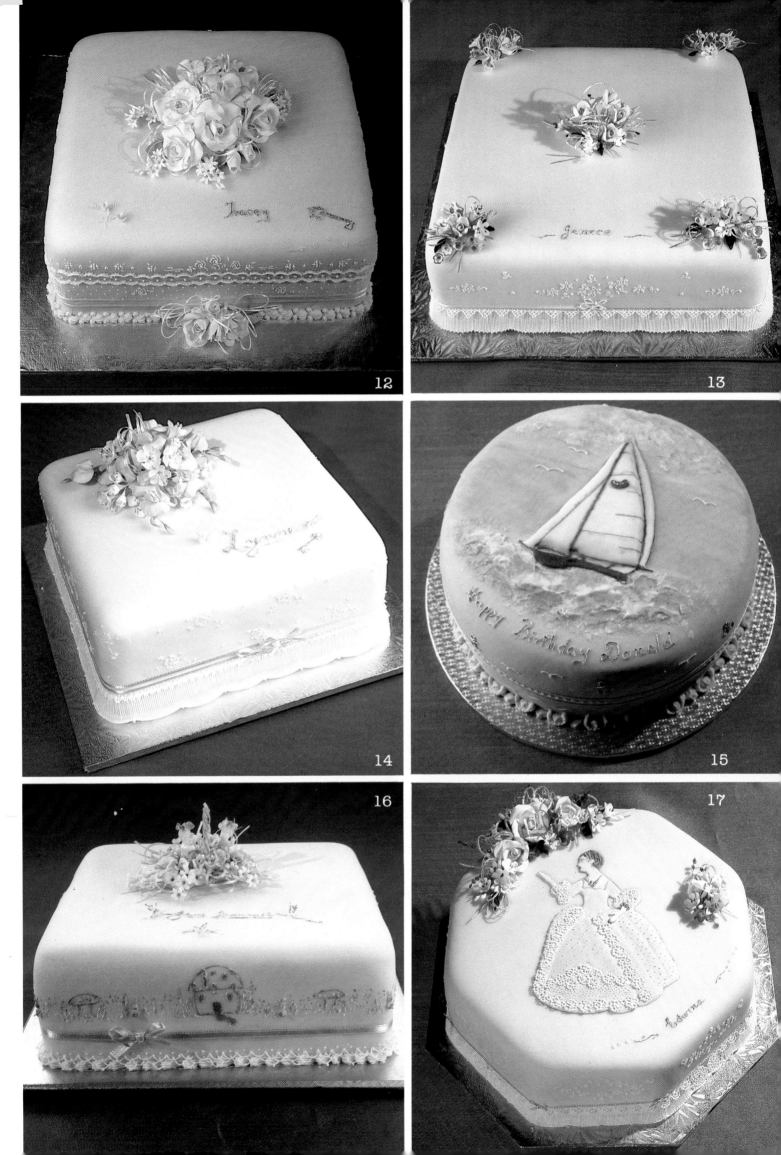

18. Browallia is a very gay flower. It too was taken from the garden and copied. The flowers vary in colour according to age, and the various sizes of buds create interesting sprays. This is a very simple flower to arrange. Crimper work is featured on the sides with hollyhock embroidery, and a shell and star base.

19. The ever-popular vintage car with bright colours is flooded directly on to the cake. Crimper work is featured, together with piped leaves and a simple shell border. The key is moulded from modelling paste and painted when dry.

20. Wildflowers which include flannel flowers, Christmas bells, wattle, Christmas bush, cocky's tongue, running postman, brown and pink boronia, and gum leaves are the feature of this cake. They are supplemented with fern embroidery, dropped loops and shell base.

21. Bauhinia, Solomon's seal and lilac make an elegant spray and a delicate colour combination on this beautiful cake. Piping is kept to a minimum to emphasise the flowers.

VIII
WEDDING CAKES

In ancient times, it was the custom of the wedding guests
to bring to the bride and groom gifts of seed — or corn,
wheat and rice. They offered the gifts as symbols
of fruitfulness and plenty. The seeds were to be planted by the
newlyweds and if the crops were fertile, so too would be
the young couple. And the two gifts, fruitfulness and
plenty, should in turn bring long life and happiness.

As the years progressed, it became uneconomical for
the guests to offer the actual seeds, so the wheat or corn
was incorporated into a special cake which custom required
to be baked by a female member of the bride's party.

Wedding cakes of today are usually toned in with
the bridemaids' colour schemes, and the favourite flowers
of the bride. They are mostly tailored and simple with
delicate colourings. Floral arrangements are carefully
positioned on the cakes in such a way as to allow
for easy cutting and to enable the bride to keep
the flowers as a memento.

This octagonal two tier
cake features Dorothy
Perkins roses, Solomon's
seal and ribbon in the
bridgework.

For this wedding cake, three tier cakes baked
in bell-shaped tins feature Dorothy Perkins
roses and buds, leaves, Solomon's seal and
snowflakes, with numerous ribbon loops.
Freehand embroidery includes pastel
coloured floodwork and fine extension
work; matching coloured split ribbon is set
into the extension work.

1. This single tier cake features a posy of daisies, forget-me-nots, blossom and a lyre. Dainty floral embroidery, split ribbon and a simple shell base with dropped loops finish the cake.

2. A very dainty two tier all-pink wedding cake features formal roses, Cecil Brunner roses and forget-me-nots. Embroidery, consisting of a scattered embossed rose and lace and fine extension work complete the design.

3. An unusual combination of a two tier cake features rectangular and heart shaped tiers with cream roses, forget-me-nots and hyacinths. Scattered embossed rose embroidery and fine extension work completes the decoration.

4. This two tier cake featuring daisies, pearl stamens in the embroidery and ribbon set into the bridgework, makes a very appealing wedding presentation.

5. Apricot coloured cattleya orchids on this two tier cake and a contrast of white blossoms, together with fine embroidery, lace and extension work are a popular choice for weddings.

6. Variation of the formal rose in apricot, using stamens, bouvardia and mauve blossoms, are featured, combined with fine embroidery, lace and extension work at the base.

7. A simple single tier cake featuring frangipani, hyacinths and snowflake, combined with dropped loops, hollyhock embroidery, split ribbon, and a simple shell and loop border.

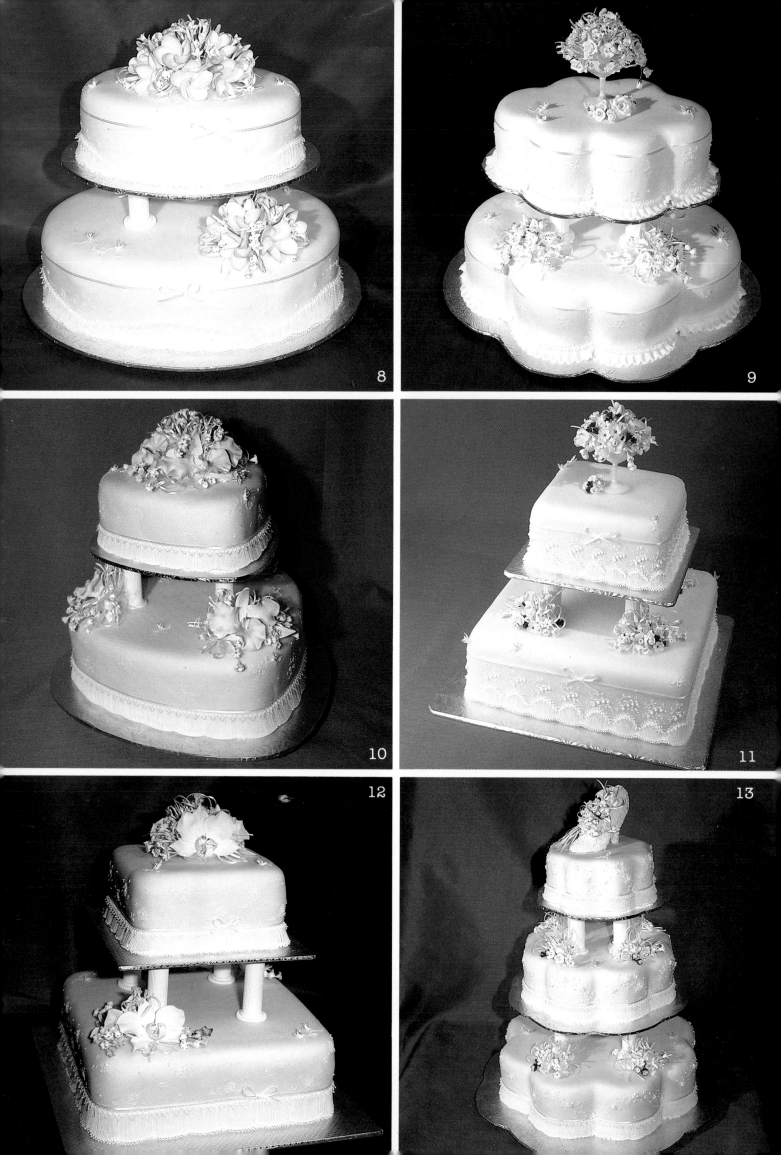

8. Pink frangipani are regal flowers. When combined with bouvardia, snowflakes and numerous ribbon loops in the posies, and fine embroidery and half-moon extension work on the sides, the result is beautiful.

9. Two posies of apricot coloured Dorothy Perkins roses, white blossom, and snowflakes are features with a frosted miniature champagne glass on the top tier of this cake. Four matching posies are used on the bottom tier. The sides of both tiers feature eyelet embroidery and a rolled fondant frill. The scalloped shape can be obtained by cutting from an oval cake.

10. Heart-shaped two tier cake features unusual pale blue cymbidiums, with blue forget-me-nots and white hyacinths. Fine scattered embroidery with tailored extension work and lace make an eye-catching cake.

11. Cecil Brunner roses on the two tier square cake, pink blossoms, hyacinths, snowdrops and a few burgundy-coloured forget-me-nots are scattered through the posies to match the bridemaids' colour scheme in this case. Vary the colour used to match the colour scheme of your wedding party. Lily of the valley embroidery and crescent-shaped extension work complete the cake.

12. White phalaenopsis orchids with pink hyacinths, snowflakes and tiny ivy leaves decorate this beautiful two tier wedding cake which also has embossed rose embroidery and fine extension work.

13. A three tier blossom shaped cake with posies of pale apricot Cecil Brunner roses, white blossom and hyacinths, a few brown boronia and snowflakes. The bridal slipper tops this creation.

14. Posies of mauve cymbidium orchids, violets and snowflakes, together with hollyhock embroidery, a simple shell border and split ribbon, make a most attractive wedding cake.

14

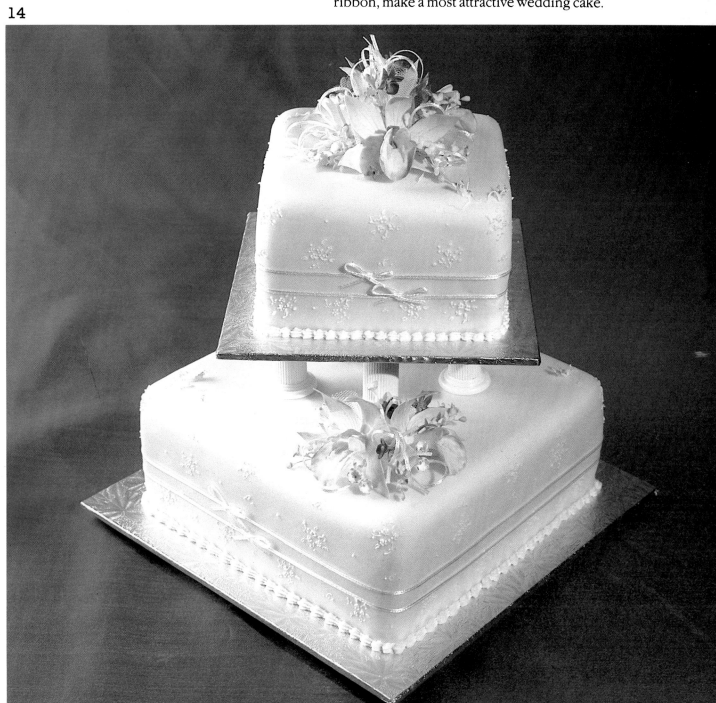

15. Two tier heart-shaped cake with full blown roses of pale pink, gypsophila and snowdrops, scattered rose embroidery with fine extension work makes a romantic wedding cake.

16. A most unusual wedding cake in milk chocolate colour with pale apricot roses and buds, white hyacinths, ribbon to match, hollyhock embroidery and a rolled fondant frill.

17. Three tier cake featuring Dorothy Perkins roses, painted red, with tiny pink gypsophila, is finished with simple embroidery and fine extension border.

18

18. This two tier heart-shaped wedding cake features regal frangipani, lily of the valley embroidery and crescent shaped extension work.

19. This one tier cake features an open shell (moulded with a real shell as model), jasmine, Dorothy Perkins roses, Cecil Brunner roses, hyacinths and forget-me-nots, snowflakes. Two wedding rings nestle inside the shell. Scattered forget-me-not embroidery, and shell with dropped loop base are also used.

20. Two tier round cake is decorated with pale pink formal roses, white hyacinths, blue forget-me-nots, Cecil Brunner roses and scattered rose embroidery. The fine extension work features loops and pulled dots at the base.

21. White carnations, Cecil Brunner roses, white hyacinths and pale blue forget-me-nots in varying sized posies are used on this cake. The bride and groom on the top complete the decoration.

22. Three tier orchid pink fan shaped wedding cake features a scalloped overlay with simple eyelet embroidery. It is decorated with white sweet peas and mauve lilac which are arranged in a small hand-made bowl on the top tier and as posies on the other tiers. Fine extension work and lace complete the decoration.

23. Three tier wedding cake features Australian wildflowers — flannel flowers, brown boronia and golden glory pea. When combined with dainty embroidery and a fondant frill they make a very impressive cake.

19

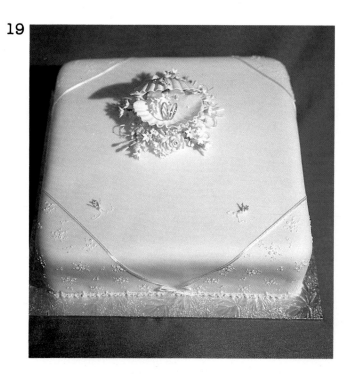

20

21

22

23

This centenary cake is decorated with
Australian wildflowers — waratah, flannel
flowers, wattle, honeyflower, wedge pea,
cocky's tongue, running postman and pink
and brown boronia. A handpainted plaque is
also featured on the top, while freehand
piped Aboriginal motifs decorate the sides,
which are finished with a simple shell border.

IX
ANNIVERSARY CAKES

Anniversaries or special occasions often call for celebrations and a cake, suitably decorated in the particular theme, can be the highlight of the event. Silver, gold, ruby and diamond wedding anniversaries are such occasions. Other celebrations, including centenary parties, school end-of-year celebrations, university and examination passes, can be made more 'special' by featuring an attractively decorated cake.

Close up of the plaque.

ANNIVERSARIES

1st year   paper

2nd year cotton

3rd year   leather

4th year   fruit and flowers

5th year   wood

6th year   sugar

7th year   wool

8th year   pottery

9th year   willow

10th year   tin

11th year   steel

12th year   silk and linen

13th year   lace

14th year   ivory

15th year   crystal

20th year   china

25th year   silver

30th year   pearl

35th year   coral

40th year   ruby

45th year   sapphire

50th year   gold

55th year   emerald

60th year   diamond

75th year   platinum

This thirtieth (pearl) wedding anniversary
cake features white roses and small flowers,
pearl embroidery and pearls incorporated
in the '30'.

This silver wedding anniversary cake
features a bird bath, open roses and violets
on the top, with a frilled rolled fondant
base and two rows of ribbons on the sides.

Twentieth wedding anniversary cake features
pink phalaenopsis and ivy leaves on the top,
with a mesh border.

59

1. This silver wedding anniversary cake features full blown roses, hollyhock embroidery and fine extension work.

2. Twenty-fifth (silver) wedding anniversary cake. It features full blown roses and small flowers in the posy with a rolled fondant frill and eyelet embroidery on the sides.

3. A ruby (fortieth) wedding anniversary cake featuring red roses, small white blossoms and hyacinths with scattered embossed rose embroidery and fine extension work.

4. This golden (fiftieth) wedding anniversary cake is decorated with deep full blown golden roses, cream Cecil Brunner roses, brown boronia and blossoms, together with simple embroidery and fine extension work.

5. Another golden wedding anniversary cake, with carnations, eyelet embroidery and a rolled fondant frill.

3

4

5

61

For this cake, the Christmas tree is shaped by
using a cookie cutter, the piped garlands
incorporate silver cachous, and the brightly
coloured parcels are made from modelling
paste, piped to resemble ribbons and bows.
Sprinkle with glitter if desired. This idea was
taken from a Christmas card.

X
CHRISTMAS CAKES

The Christmas cake is a bright and colourful centrepiece
for the festive table. It also makes a beautiful gift for
the person who has everything — a personal touch
is something that money cannot buy. Poinsettia, Christmas
rose, holly berries and leaves, sugar and moulded
bells, flooded Father Christmas are
illustrated in the following pages. A favourite
Christmas card can be a source of inspiration, as can
Christmas wrapping paper and so on.

The red poinsettia lends its gay colour to
many things at Christmas time. Its vividness,
together with the unusual shape of the cake,
and the contrast of the holly embroidery,
gives an eye-catching result.

This design for Santa and his reindeer was
taken from a Christmas card. The plaque was
flooded before being placed on the cake.
Roofs were flooded on to the waxed paper
placed over a tin for shaping, allowed to dry
and then put on to the sides of the cake.

1. This cake, featuring Christmas roses, pine cones, holly embroidery, with a star and loop base, was inspired by a Christmas card.

2. The Father Christmas boot is moulded in modelling paste, flooded and filled with presents made of modelling paste, surrounded with moulded holly. The piping on the sides is also of holly and stocking design.

3. A very simple wildflower design for a Christmas cake consisting of Sturt's desert pea, wattle, honey flowers, pink and brown boronia and flannel flowers.

4. Cream and red poinsettia, candles and Christmas bells adorn this elegant cake. Crimper work and ribbon insertion look attractive and may give you further ideas.

5. Red candles, Christmas rose, holly and simple snowflake embroidery are the highlights of this cake. The base incorporates petal and star pipes.

6. A Tyrolean girl in a field of daisies was inspired by a European Christmas card. The sides feature flooded Christmas trees. There are moulded Christmas roses at the corners on the base of the cake.

7. Father Christmas was baked in a rectangular shaped tin, then cut to shape. The details were built up in rolled fondant, then finished with piping and flooding. The chimney also forms part of the cake. The holly and parcel are moulded.

Humpty Dumpty cake. This cake can be
made from a butter mixture, such as a marble
cake, or a fruit cake. Bake the wall in a
rectangular shaped tin or cut to shape. Cover
with rolled fondant and, while still soft, mark
with the back of a knife to resemble bricks.
Humpty Dumpty was made of modelling
paste, set in a two piece egg mould which is
joined with royal icing. Small details were
then painted, and the arms, hat, tie, legs
were moulded from modelling paste.

XI
NOVELTY CAKES

Novelty cakes create special interest for children.
There are a variety of doll cakes, including a bride doll,
Humpty Dumpty, a doll's house, a tepee and so on.

Novelty cakes can also be piped simply to be effective;
colour and colour blending is an important factor
in decorating.

Bride doll. The bride cake was baked in a
Dolly Varden tin, and features coloured
eyelet embroidery, a full circle flounced skirt,
veil, petal pipe frill at neckline and a bouquet
of tiny flowers.

1. Queen of Hearts doll. The cake was baked in a Dolly Varden tin. The crown is made from gold-painted cardboard, the red hearts are stamped out of modelling paste with a small cutter. She carries a small moulded tray with piped tarts.

2. A Tyrolean doll complete with an embroidered skirt, plain white apron and a posy of tiny flowers.

3. A very simple doll cake with a variation of the rolled fondant frill at the neckline and on the hat. The skirt is finished with a fondant overlay.

4. Dolly Varden cake features a completely piped decoration. Always remember to add 4-5 drops of glycerine to each egg white to keep the royal icing from becoming too hard.

5. This elegant doll cake features embroidery (showing continental influence) which has been flooded. The underskirts are of rolled fondant shaped into frills. The lady holds a posy of red roses.

5

6

6. This Confirmation cake was baked in a rectangular shaped tin and cut. A combination of two rectangular tins also gives the same effect. Red roses and tiny white blossoms adorn the cake, as do cornelli, crimper and ribbon insertion.

7. Double birthday cake. Two large cakes were cut to shape for the double event. Flooded motifs were taken from nursery wrapping paper. The edges of leaves were tipped with gold and the bookmark is of variegated ribbon.

8. Log cake. Swiss roll covered with Vienna icing, shredded green coloured coconut is used for grass. Animals, mushrooms and ladder are moulded from modelling paste, and vine and leaves piped from royal icing. This cake may be made as simple or elaborate as desired.

9. Guitar cake is baked in a rectangular tin and cut to shape. This is very popular with teenagers. Coloured medallions on the sides were made out of modelling paste, the strings are wire and are attached to modelling paste painted silver.

10. Book cakes are baked in a special open book tin. Decorations are piped daisies, flooded leaves, and handpainted motifs of mushroom and snails. The edges are piped with the basket weave tube to resemble the leaves of a book then, when dry, painted with non-toxic gold paint.

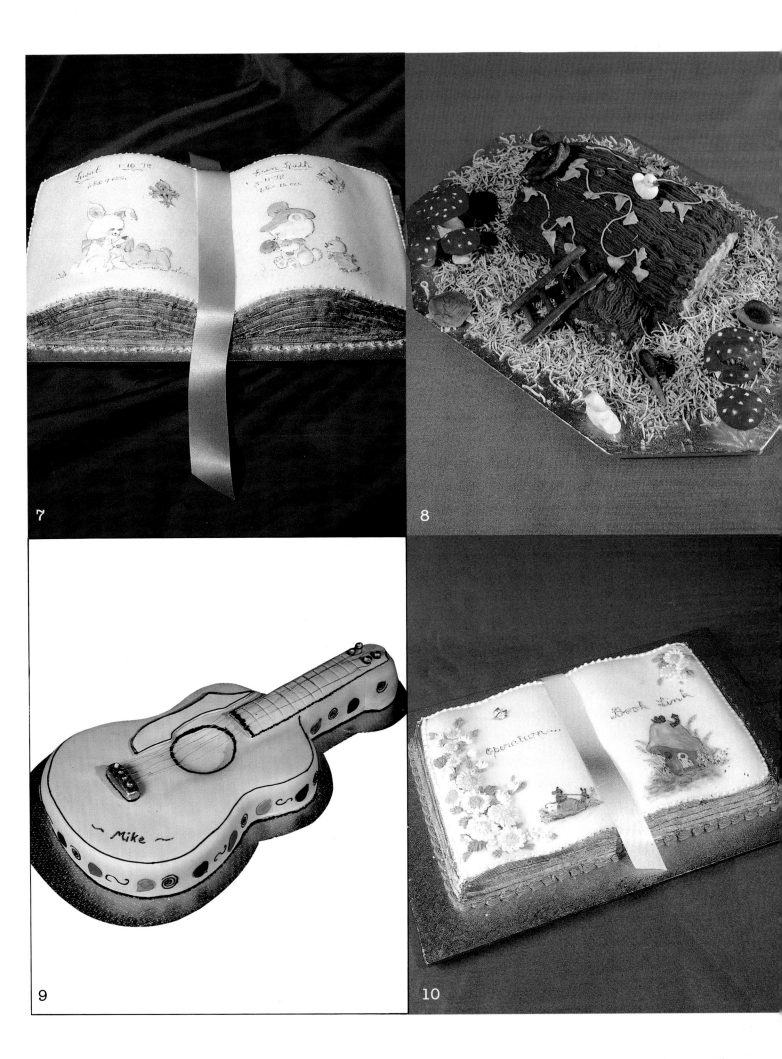

7

8

9

10

Elves' house. The cake is baked in a small oval tin and then covered with rolled fondant. Thatched roof is made with almond paste mixed with dry cocoa or drinking chocolate. Piped wisteria over the roof and piped trees (over a cardboard cone) add colour. The door, window frames and toadstools were made of modelling paste. There are small piped flowers and leaves in the front garden, with a path of brown sugar, grass (green coconut); the fairies are tiny dolls with piped skirts.

The teepee is baked in a Dolly Varden tin.
Satay skewers are used for poles, the rolled
fondant rolled out to a semi-circle, draped
around the poles and secured. Decorations
are hand-painted with diluted coloured
royal icing. Arrange toy cowboys and
Indians on a green coconut base.

Log cabin. Cut the cake to shape. Using No
5 star tube (with 4–5 drops of glycerine
added to each egg white) pipe the front and
sides of the house. Cut a piece of heavy
cardboard for the roof, bend to shape and
adhere with royal icing. Pipe the roof, attach
chimney (moulded or made from cardboard) and pipe over the chimney. Add
cotton wool to the chimney for smoke. Pipe
wisteria and leaves over the roof. Flood
coloured pathway, add flowerbox (moulded
in modelling paste), place toothpick flowers
in the box. Use green coconut for grass.

Tyrolean doll's house. Bake a large cake,
either fruit or heavy butter cake, and cut to
shape. Cover with cream rolled fondant
icing. The roof is moulded from pastillage
(see p. 12) rolled and cut in two pieces and
set with royal icing in place. Tiles are cut
with rose petal cutters. Shutters are also
moulded from pastillage. Small red flowers
and leaves are placed in the windowboxes.

This cake features moulded New Zealand wildflowers kowhai, pohutakawa, kaka beak and clematis. A flooded fantail bird and kiwi are featured on the top, with piped Maori symbols on the sides.

78

XII
MISCELLANEOUS

A plaque of phalaenopsis orchids, hyacinths
and snowflake.

Field flower cake of brightly coloured poppies, cornflowers, daisies, buttercups and wheat, with a small matching posy on the board, is very popular for a cake with a difference.

A close up of the flowers.

Bon Voyage cake displaying Australian wildflowers — waratah, flannel flowers, eggs and bacon, and gumnuts — and boomerangs. Embroidery features piped boomerangs and Aboriginal motifs.

This Bon Voyage cake could also be used for a boy's birthday cake. It features wattle, flannel flowers, pink and brown boronia in the posies, and has simple shell border.

5

1. Combination twenty-first and engagement cake. One corner features pink roses for the girl, the other blue for the boy. A shell base and simple fleur de lis embroidery complete an unusual cake.

2. A small vase filled with popular Java orchids. This type of decoration is suitable for the top of a wedding cake.

3. Piped flowers — pansies, forget-me-nots — arranged on a flooded horseshoe, make an ideal cake for men. This one has a simple puff border.

4. A Mother's Day cake with yellow carnations and wild violets in the posy. A frilled rolled fondant overlay with ribbon insertion with soft embroidery, is draped to show the fine extension work at the base.

5. A child's handkerchief was traced and the designs transferred to the cake, then flooded in coloured icing directly on to the cake. The sides are painted to resemble grass and moulded mushrooms are placed at the corners.

An octagonal blossom-tin shaped cake
features a large spray of blossoms,
forget-me-nots, Cecil Brunner roses and
hyacinths together with hollyhock
embroidery and piped on lace. This makes a
very elegant engagement cake.

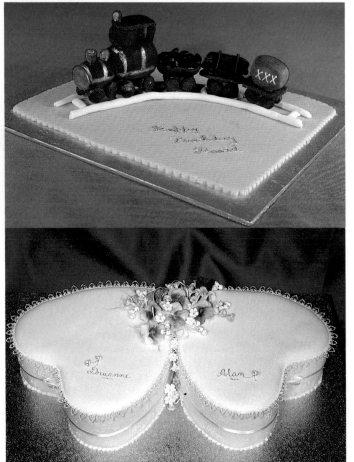

This plaque features a moulded train made
from rolled covering fondant.

Twin hearts engagement cake featuring lace
and simple scroll embroidery; pansies form
the spray.

A chocolate Easter egg decorated with
New Zealand wildflowers — manuka,
Chatham Island forget-me-not, clematis.

This egg cake was baked in two halves in a special tin. It may also be used for a football cake. Floodwork is the feature of this cake: a colourful fox, holding balloons. A rolled fondant frill completes the decoration.

Decorated eggs are very popular at Easter. Simple decorations of flowers and ribbons can transform a plain candy or chocolate egg into a delightful gift.

The pink egg is made from crystal sugar and egg white. The sugar is moistened with egg white to a wet-sand consistency and set in two plastic moulds. These moulds are available in specialist stores just prior to Easter.

The other eggs are made from modelling paste. Windows were cut while the rolled fondant was still soft and the eggs were decorated with blossoms of pink, mauve and white daisies.

Corn poppy

Buttercup

Wheat

Field daisy

Cornflower

STEP-BY-STEP
PHOTOGRAPHS

Solomon's seal

Lilac

White bauhinia

Pansy and viola

Geranium

Ifala lily

Easter daisy

Tuber rose

Narcissus

Daffodil

Wood violet

Ground orchid

Iceberg rose

Poinsettia

Java orchid

Waterlily

Browallia

Apricot or quince blossom

Jasmine

Beloperone or shrimp plant

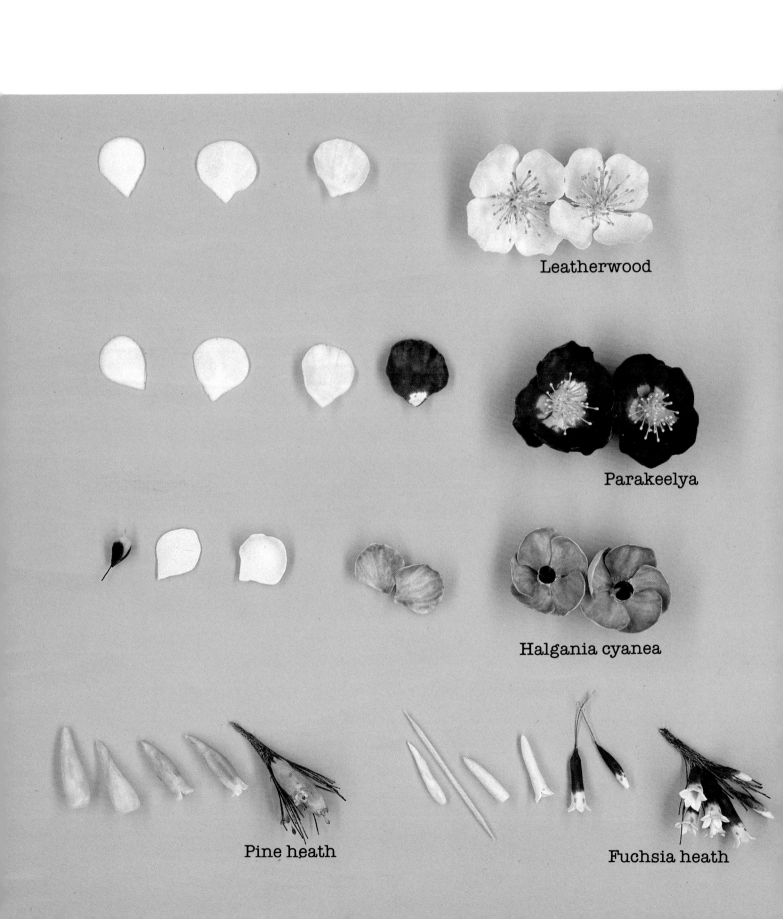

Leatherwood

Parakeelya

Halgania cyanea

Pine heath

Fuchsia heath

Common correa

Native fuchsia

Wild violets

Pine cones

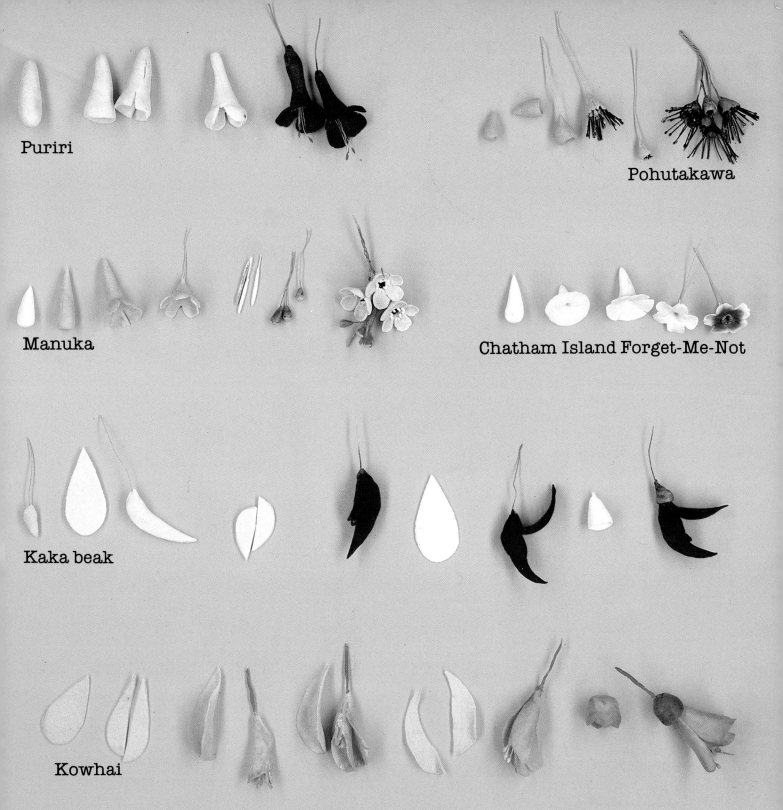

Puriri

Pohutakawa

Manuka

Chatham Island Forget-Me-Not

Kaka beak

Kowhai

95

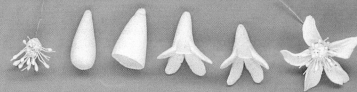

Clematis

Mountain lacebark

New Zealand eyebright

Mount Earnslaw ourisia

Poroporo

Kotukutuku

PATTERNS

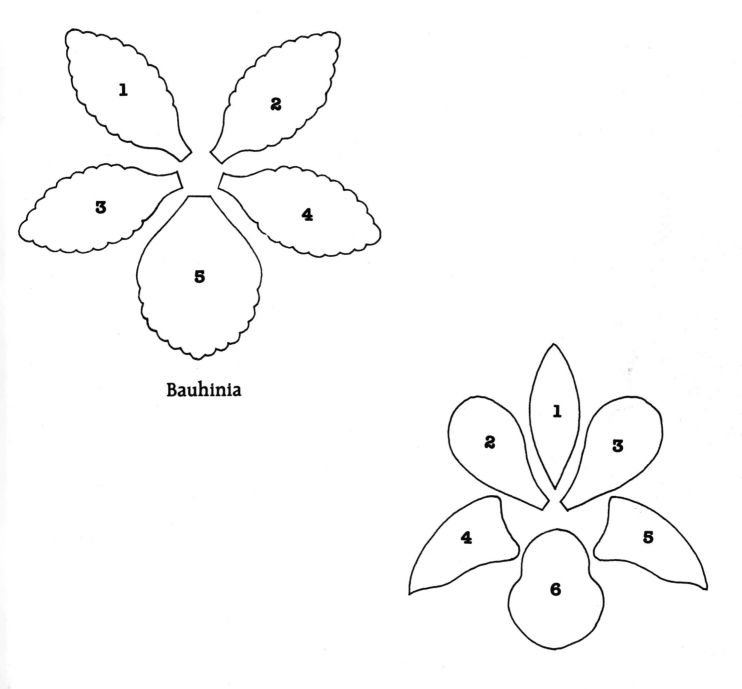

Bauhinia

Java Orchid

Daffodil   Jonquil

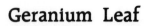

Geranium Leaf

Corn Poppy   Buttercup

Field-flower Leaf

Geranium   Pansy   Viola

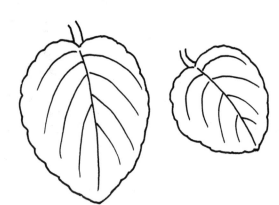

Leatherwood   Halgania   Parakeela

Violet Leaf

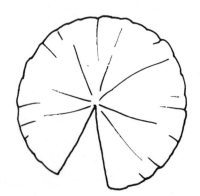

 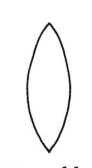 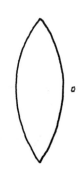

Water-lily   Water-lily

Kiwi

Kingfisher

Cockatoo

Kookaburra

Koala

Fantail

Platypus

Wagtail

Blue Wren

Aquarius

Pisces

Aries

Taurus

Gemini

Cancer

Leo

Virgo

Libra

Scorpio

Sagittarius

Capricorn

A B C D E F G

H I J K L M N

O P Q R S T U

V W X Y Z

a b c d e f g h i j k l m

n o p q r s t u v w x y z

Congratulations

Best Wishes

Best Wishes   Congratulations

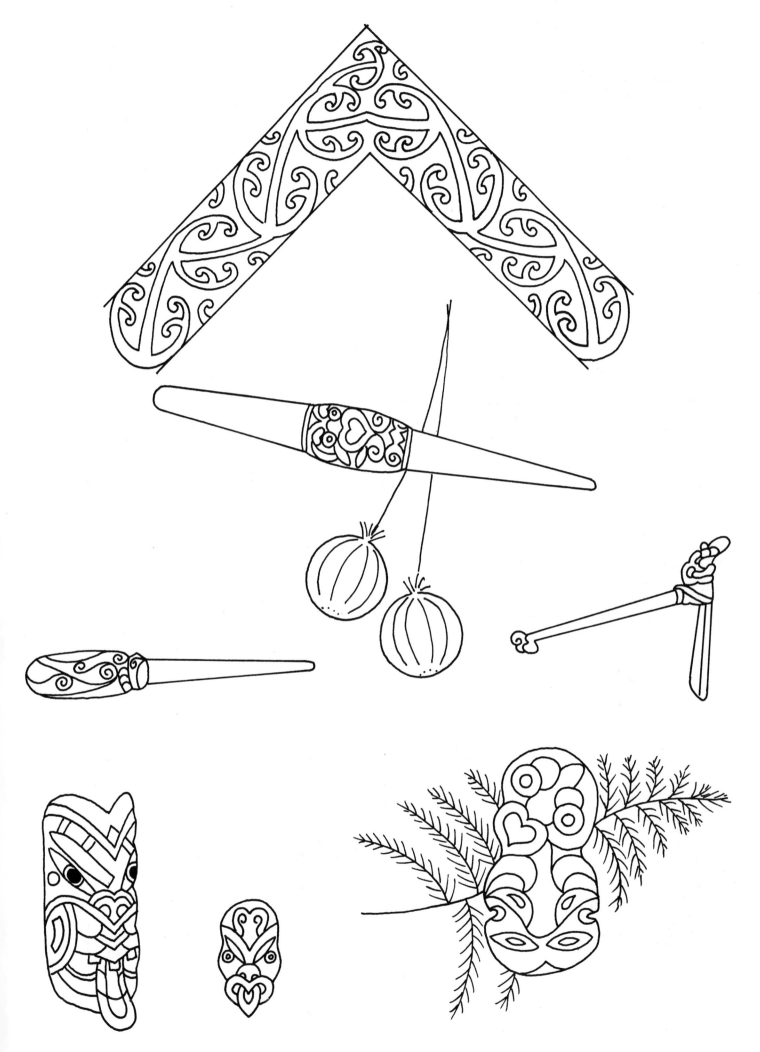

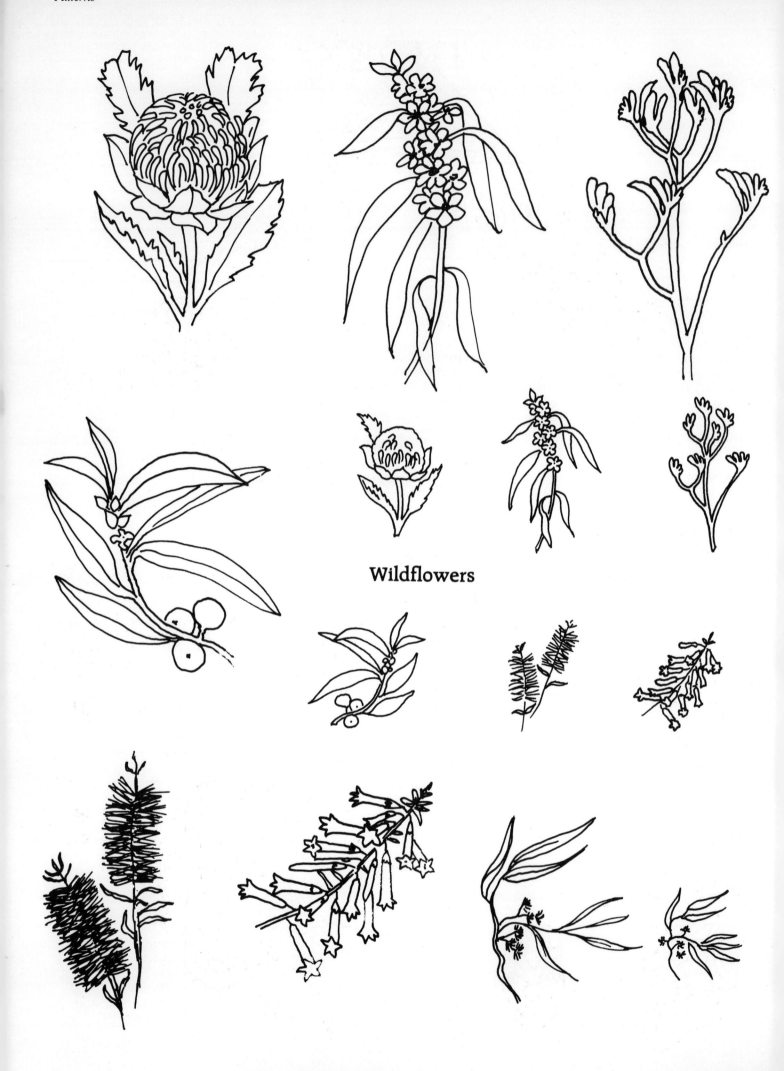

Wildflowers

INDEX

Almond paste (recipe)   11
Anniversary cakes   57-62
Apricot blossom   26, 92

Bauhinia, White   23, 89
Beloperone   27, 92
Birthday cakes   39-45
Blossom cake   84
Bluebirds   19
Bon voyage cake   81
Book cake   72
Browallia   26, 92
Butter cake, large rich (recipe)   13
Butter icing (recipe)   13
Buttercup   21, 88

Cakes, anniversary   57-62
Cakes, birthday   39-45
Cakes, Christening   35-38
Cakes, Christmas   63-67
Cakes, doll   69-70
Cakes, miscellaneous   78-87
Cakes, novelty   68-78
Cakes, wedding   47-56
Cakes see also Recipes
Clematis   32, 96
Chatham Island forget-me-not   32, 95
Child's handkerchief cake   83
Christening cakes   35-38
Christmas cakes   63-67
Colourings   20
Cone, hollow basic   15
Cones, pine   31, 94
Confirmation cake   72
Corn poppy   21, 88
Cornflower   22, 88
Correa, common   30, 94

Daffodil   24, 90
Daisy, Easter   24, 90
Daisy, field   22, 88
Doll cakes   69-70
Dropped string work   17-19

Easter egg, chocolate   84
Easter eggs, decorated   86
Easter daisy   24, 90
Egg cake   86
Eggs, decorated   86
Elves' house cake   74
Embroidery   16
Embroidery, eyelet   15
Equipment   8-9
Extension work   17-19
Eyebright, New Zealand   33, 96

Ferns   20
Field daisy   22, 88
Field flower guide   80
Flag method   20
Floodwork variations   15
Floral plaque   79
Flowers field   21
Flowers moulded   21-27
Flower paste (recipe)   13
Forget-me-not, Chatham Island   32, 95
Fruit cake, rich (recipe)   12
Fuschia heath   30, 93
Fuschia, native   30, 94
Fuschia, New Zealand native   34, 96

Genoise cake (recipe)   13
Geranium   23, 89
Ground orchid   25, 91
Guitar cake   72
Gum paste (recipe)   12

Halgania cyanea   29, 93
Hearts, twin, cake   84
Heath, fuschia   30, 93
Heath, pine   30, 93
Horseshoe cake   83
Humpty Dumpty cake   68

Iceberg rose   25, 91
Icings see Recipes
Ifala lily   23, 90

Jasmine   26, 92
Java orchid   25, 92

Kaka beak   32, 95
Kotukutuku   34, 96
Kowhai   32, 95

Lacebark, mountain   33, 96
Leatherwood   29, 93
Lilac   22, 89
Log cabin cake   76
Log cake   72

Manuka   32, 95
Marble cake (recipe)   13
Marzipan (recipe)   11
Miscellaneous cakes   78-87
Mothers' Day cake   83
Moulded flowers   21-27
Mount Earnslaw ourisia   33, 96
Mountain lacebark   33,96

Narcissus   24, 90

New Zealand nature cake   78
Novelty cakes   68-78

Orchid ground   25, 91
Orchid, Java   25, 92
Ourisia, Mount Earnslaw   33, 96

Pansies   23, 89
Parakeelya   29, 93
Paste, modelling (recipe)   12
Pastillage (recipe)   12
Patterns   97-108
Pea base   15
Pine cones   31, 94
Pine heath   30, 93
Plaque, floral   79
Pohutakawa   31, 95
Poinsettia   25, 91
Poppy, corn   21, 88
Poroporo   33, 96
Posy arranging   20
Puriri   31, 95

Quince blossom   26, 92

Recipes   11-13
Rolled fondant (recipe)   11
Rolled fondant frill   15
Rolled fondant overlay   15
Rolled fondant, plastic (recipe)   12
Rose, iceberg   25, 91
Rose, tuber   24, 90
Royal icing (recipe)   11

Shrimp plant   27, 92
Solomon's seal   22, 89

Techniques, supplementary   15-20
Teepee cake   75
Train plaque   84
Tuber rose   24, 90
Tyrolese doll's house cake   76

Vienna icing (recipe)   13
Violas   23, 89
Violets, wild   31, 94
Violets, wood   24, 91

Warm Icing (recipe)   12
Waterlily   26, 92
Wedding cakes   47-56
Wheat   22, 88
Wildflowers, Australian   29-31
Wildflowers, New Zealand   31-34
Wood violet   24-91